Great Western Castle Class 4-6-0 Locomotives

The Final Years 1960–1965

Front cover: 4080 *Powderham Castle* at its home depot of Cardiff Canton, 17 August 1962. Although built in 1924, it was rebuilt as late as August 1958 with 4-row superheater boiler and double chimney. It was withdrawn in August 1964 after accumulating the highest mileage in traffic (1,974,461) of any Castle, other than the combined mileage of the rebuilt 'Stars' in both forms. (Alan Jarvis/SLS)

Back cover photos:
Reading's double-chimneyed 4074 *Caldicot Castle* climbing Hatton bank with little apparent effort on the 9.25am Margate – Wolverhampton, August 1962. (Derek Penney)

Laira's 4087 *Cardigan Castle* with new front end section, double-chimney, 4-row superheat boiler and valveless mechanical lubricator at its home depot, 27 January 1962. (Peter Gray/GW Trust)

4079 *Pendennis Castle* with the high speed special, *The Great Western,* Z48 of 9 May 1964, between Reading West and Southcote Junction. (Derek Penney)

Great Western Castle Class 4-6-0 Locomotives

The Final Years 1960–1965

DAVID MAIDMENT

AN IMPRINT OF PEN & SWORD BOOKS LTD.
YORKSHIRE – PHILADELPHIA

First published in Great Britain in 2022 by
Pen & Sword Transport
An imprint of Pen & Sword Books Ltd
Yorkshire - Philadelphia

Copyright © David Maidment, 2022

ISBN 978 1 39909 534 1

The right of David Maidment to be identified as author of this work has been asserted by him in accordance with the Copyright, Designs and Patents Act 1988.

A CIP catalogue record for this book is available from the British Library.

All rights reserved. No part of this book may be reproduced or transmitted in any form or by any means, electronic or mechanical including photocopying, recording or by any information storage and retrieval system, without permission from the Publisher in writing.

Typeset in Palatino by SJmagic DESIGN SERVICES, India.
Printed and bound in India by Replika Press Pvt. Ltd.

Pen & Sword Books Ltd incorporates the Imprints of Pen & Sword Books Archaeology, Atlas, Aviation, Battleground, Discovery, Family History, History, Maritime, Military, Naval, Politics, Railways, Select, Transport, True Crime, Fiction, Frontline Books, Leo Cooper, Praetorian Press, Seaforth Publishing, Wharncliffe and White Owl.

For a complete list of Pen & Sword titles please contact:

PEN & SWORD BOOKS LIMITED
47 Church Street, Barnsley, South Yorkshire, S70 2AS, England
E-mail: enquiries@pen-and-sword.co.uk
Website: www.pen-and-sword.co.uk

Or

PEN AND SWORD BOOKS
1950 Lawrence Rd, Havertown, PA 19083, USA
E-mail: Uspen-and-sword@casematepublishers.com
Website: www.penandswordbooks.com

All royalties from this book will be donated to the Railway Children charity [reg. no. 1058991] [www.railwaychildren.org.uk]

Other books by David Maidment:

Novels (Religious historical fiction)
The Child Madonna, Melrose Books, 2009
The Missing Madonna, PublishNation, 2012
The Madonna and her Sons, PublishNation, 2015
The Reluctant Traitor, PublishNation, 2021

Novels (Railway fiction)
Lives on the Line, Max Books, 2013
Steamy Stories, PublishNation, 2021 (Short stories)

Non-fiction (Railways)
The Toss of a Coin, PublishNation, 2014
A Privileged Journey, Pen & Sword, 2015
An Indian Summer of Steam, Pen & Sword, 2015
Great Western Eight-Coupled Heavy Freight Locomotives, Pen & Sword, 2015
Great Western Moguls and Prairies, Pen & Sword, 2016
Southern Urie and Maunsell 2-cylinder 4-6-0s, Pen & Sword, 2016
Great Western Small-Wheeled Double-Framed 4-4-0s, Pen & Sword, 2017
The Development of the German Pacific Locomotive, Pen & Sword, 2017
Great Western Large-Wheeled Double-Framed 4-4-0s, Pen & Sword, 2017
Great Western Counties, 4-4-0s, 4-4-2Ts & 4-6-0s, Pen & Sword, 2018
Southern Maunsell Moguls and Tank Engines, Pen & Sword, 2018
Southern Maunsell 4-4-0s, Pen & Sword, 2019
Great Western Granges, Pen & Sword, 2019
Cambrian Railways Gallery, with Paul Carpenter, Pen & Sword, 2019
Great Western Panniers, Pen & Sword, 2019
Great Western Kings, Pen & Sword, 2020
Great Western & Absorbed Railway 0-6-2Ts, Pen & Sword, 2020
Drummond's L&SWR Passenger & Mixed Traffic Locomotives, Pen & Sword, 2020
Southern 0-6-0 Tender Locomotives, Pen & Sword, 2021
LNER 4-6-0 Locomotives, Pen & Sword, 2021
Midland & LMS 4-4-0s, Pen & Sword, 2021
Great Western Castle 4-6-0 Locomotives, 1923-1959, with Bob Meanley, Pen & Sword, 2022

Non-fiction (Street Children)
The Other Railway Children, PublishNation, 2012
Nobody ever listened to me, PublishNation, 2012

CONTENTS

Preface & Acknowledgements 6

Chapter 1 Recap – The First Thirty-Five Years 8

Chapter 2 The Last Years at the Top – 1960–1962 17

Chapter 3 Personal Recollections – 1960–1963 51

Chapter 4 The End Draws Near – 1963–1965 76

Chapter 5 Conclusions 110

Colour Section 113

Appendix 169

Bibliography 177

Index 178

PREFACE & ACKNOWLEDGEMENTS

This is the second of the three books I really wanted to write. I remember standing on Bristol Temple Meads station in the winter of 1944 and, after being evacuated with my mother and toddler sister, seeing my journey home start on a train hauled by 4087 *Cardigan Castle*. When, like many young boys of that era, I became a trainspotter in 1947, the first number underlined in my new GWR Ian Allan ABC book was that of 4087. You will see that several photographs of that engine have sneaked their way into this book! My love of the 'Castles' was further embedded in me during five years of vacation work and railway training at Old Oak Common between 1957 and 1962.

This book covers the final years from 1960 to the withdrawal of 7029 *Clun Castle* at the end of 1965. It relates the history, operation and performance of the 'Castles', and allows me to include my own extensive personal experiences of the engines, particularly that between 1960 and 1964 when my work brought me into direct contact with many of the class.

I owe much to many – those who have trodden this path before and whose books I researched and are acknowledged in the bibliography at the end of the book. To both Bob Meanley and John Hodge who reviewed and commented on the text and John provided many of the photographs taken in South Wales of both his own and his splendid collections of Alan Jarvis and R.K. Davies. To Steve Bartlett and Brian Penney for access to their research and articles. To Derek Penney, Graham Stacey, Brian Stephenson and the Rail Archive Stephenson, Paul Shackcloth and the Manchester Locomotive Society for access to their vast collection of photographs and for allowing me to use them free of any publication fee and Laurence Waters of the Great Western Trust for many of the colour photos from their collection as I'm donating 'as is my custom' all the royalties to the Railway Children charity (www.railwaychildren.org.uk) which I founded in 1995 and which supports street and runaway children picked up on railway and other transport terminals of the world – at the current time in India, East Africa and the United Kingdom. I have tried to trace and contact the copyright holders of all the photographs but if I have missed anyone, please get in touch with the publisher so I can make amends.

I'm also grateful to John Scott-Morgan, friend and Commissioning Editor of Pen and Sword, Carol Trow my editor and Janet Brookes and the Pen and Sword design, production and marketing team for their encouragement, support and professionalism. I commend the book to all those who like myself had a special soft spot for these engines, and to those who like to model particular examples as, at least during the last decade of their existence, there were so many varieties and differences among the last survivors – some individual engines managed in the last few years of the class to sport at one time or another 'joggled' frames, straight frames with dished space for the bogie wheels, renewed straight frame front sections, two, three and four-row superheaters, single chimneys, tall and short, double chimneys, outside slim steam pipes as built and the later chunkier steam pipes, hydrostatic lubricators, mechanical lubricators in front or behind the outside steampipe or even halfway up the side of the smokebox, Collett tenders or Hawksworth tenders. My favourite 4087 managed to embrace every single one of these variations during its 38-year career! When I was virtually 'living' at Old Oak Common in 1957, 1958 and 1962, I reckoned I could tell the identity of most of the Castles coming on shed before they were near enough to read the number!

So here they are at the start of this book in 1960, just 162 of the original 171 now, with only 4037 left of the early 'Star' rebuilds. I saw every one except 100A1 *Lloyds* which was withdrawn before my first day's trainspotting round the London termini. As I recounted in my previous book, I travelled behind 156 of them, 25,000 miles, 36 times behind 5043, not all of them since preservation, 15 times behind 7029, 14 behind 5025, 12 behind 5039. I have ridden on the footplate of 28 different Castles, some more than once, for nearly 2,000 miles. I've been inside the warm firebox of one with Old Oak's boilersmith, I maintained the record cards for some 35 Old Oak Castle residents for six months, I have proposed them for Swindon Works overhaul, I believe I have sufficient authority to complete their story, as most of that experience was in the years from 1957 onwards.

David Maidment
November 2022

Chapter 1
RECAP – THE FIRST THIRTY-FIVE YEARS

George Jackson Churchward was an outstanding mechanical engineer. Not necessarily a great inventor, he had the gift of being aware of developments in railway locomotive design worldwide, recognising their significance and bringing together all the best from American, Continental and British experience to produce a family of standardised locomotive types to cover all the needs of the Great Western Railway some ten to fifteen years ahead of his contemporaries. His 4-6-0 express passenger locomotives, his heavy freight 2-8-0s and his large and small 'Prairie' and 2-8-0 tank engines formed the bedrock, meeting the company's needs well into the 1920s and were the basis on which his successor, Charles Collett, was able to build between the two world wars.

The Castles, basically enlarged 'Stars', took the railway world by storm in the mid-1920s, and influenced the mechanical engineering policies of the other three British railway companies, and became one of the longest production lines, Hawksworth continuing their construction at Swindon into the British Railways days of 1950, their building spread over twenty-seven years. Their impact in hauling prodigious loads from London to Devon at economical rates far surpassing that of other railways' locomotives in the 1920s was sensational and hardly believed by Collett's contemporaries. Their high-speed records in the 1930s ushered in an era that was only surpassed by Stanier's and Gresley's streamliners. Collett did not attempt to improve on these as his fleet met all his own company's needs in times of stringent economy.

The new locomotives performed well 'straight from the box', having few teething problems and requiring little need for design modification until the post-war period. A straightening of the front section of the main frame after the first twenty, the provision of 4,000 gallon in place of the former 3,500 gallon tenders, and a slight reduction in firebox dimensions to enable greater space for boiler washouts were the only significant changes in their first twenty years. After the Second World War, the need for greater superheat to combat the reduced access to good steam coal and the consequent need for mechanical lubrication was recognised by Hawksworth and incorporated in the forty Castles built between 1946 and 1950. 5049 was provided with a larger 4-row superheater in 1949. Swindon's research team led by Sam Ell carried out a series of tests on exhaust draughting in the early 1950s, culminating in improved draughting for the single chimneyed engines and then the provision of double chimneys and 4-row superheater boilers fitted to sixty-six of the Castles between 1956 and 1961. With the '5098' class as built in 1946, tests revealed that 1,800hp could be produced at 80mph using full regulator and 25 per cent cut-off. A double-chimney Castle was tested at 20 per cent cut-off at 86mph and recorded a pull of 3,600lbs.

The improvements led to acceleration of the Western Region's expresses from 1954 and the restoration of a few 60mph+ start-to-stop schedules of key Castle-hauled named trains like the *Bristolian, Cheltenham Spa Express, Pembroke Coast Express, Cambrian Coast Express* and the

Cathedrals Express. The 1955 Railway Modernisation Plan identified the replacement of steam traction by some dieselisation as a stop gap to eventual electrification, but its impact was not expected to affect the majority of the Castles' work until the mid to late 1960s, with initially the diesel hydraulics impacting the West of England services between 1959 and 1962. Then the deteriorating BR finances and new management on the Region from 1962 accelerated the change and many of the improvements to the Castles (and Kings) proved shortlived as mass dieselisation took place on the Region between 1962 and 1965. The Kings retained their top link work on the Birmingham-Wolverhampton road to their end and the Castles remained on top link work to Worcester, Hereford and Gloucester until the end of 1963 and were only relegated to diesel standby duties, parcels, freight and summer Saturday working in their final two years.

The first volume describing the design, construction, operation and performance of the Castles in their first thirty-five years was entitled *The Great Western Castle 4-6-0 Locomotives, 1923-1959* and was published by Pen and Sword earlier in 2022. This volume will continue their story in their final years, when from 1960-62 they still held sway on all the main WR routes except the Paddington-Bristol/Plymouth lines. Then after dieselisation of the South Wales and Birmingham main lines in the summer of 1962 and the significant withdrawal of Castles at the end of that summer timetable, their diminishing role is recognised but the remaining engines were still capable of high performance when the situation demanded it. Eight 'Castles' were preserved, six of them restored for operation at times between 1965 and the twenty-first century and their story will be told in a third volume about the class.

For the sake of completeness, some of the key statistics about the lives of the individual class members are repeated in the appendix to this book as well as appearing in a fuller form in the first volume.

Churchward 4-cylinder 'Star' 4051 *Princess Helena*, designed in 1907 and constructed in 1914, at Old Oak Common, c1923. Churchward had wished to put the larger Swindon No.7 boiler on a Star frame but was thwarted by axleweight restrictions placed by the railway's Chief Civil Engineer. (Manchester Locomotive Society (MLS) Collection)

Collett's 4079
Pendennis Castle as constructed new in 1924 with the front end of the frame 'joggled' to clear the bogie wheels and as first fitted with a 3,500 gallon tender. 4079 was chosen to represent the class in the 1925 locomotive exchange with a Gresley Pacific on the LNER main line from King's Cross to Leeds.
(F. Moore/MLS Collection)

4082 *Windsor Castle*, the last of the first ten constructed in 1924, in original condition apart from the provision of a 4,000 gallon tender, working a Paddington-Barry Island excursion train near Barry Town station, c1934.
(J.G. Hubback/John Hodge Collection)

Collett rebuilt the GWR's lone Pacific, Churchward's 111 *The Great Bear*, as a 4-6-0 Castle in 1924 and after displaying it at the Darlington Railway Centenary Exhibition in 1925, allocated it to Old Oak Common. It is seen here heading a Swindon Works staff outing to Barry Island in the mid-1930s passing Cadoxton yard. In addition to 111, Collett rebuilt five early 'Stars' as Castles in the 1920s – 4000/09/16/32/37. (J.G. Hubback/John Hodge Collection)

5050 *Devizes* Castle seen here as constructed in 1936 and before renaming as *Earl of St Germans* in 1937. It was part of the '5013' series built from 1932 onwards with straight frames dished to clear the bogie wheels, a rectangular inside cylinder block casing, shorter single chimney and container for the long fire irons behind the rear splasher. (MLS Collection)

5005 Manorbier Castle was partially streamlined in 1935 at the GWR Board's instigation to match in publicity terms the streamlining developments on the LMS and LNER. Initially streamlining was applied from the buffer beam to enclose the cylinders and steam pipes, but this was quickly removed to allow air to flow when working to avoid overheating of the motion and axleboxes. 5005 is seen here in Swindon shed with the air-smoothing around the smokebox, chimney, firebox, nameplate and cab. All was removed during the Second World War. (MLS Collection)

Collett decided to rebuild 'Stars' of the 1922/3 'Abbey' series from 1937 and the first example chosen was 4064 *Reading Abbey* seen here in the 1920s before rebuilding. It was renumbered 5084, but retained its name. 4063-4072 were rebuilt as 5083-5092 by 1941, leaving only 4061 and 4062 of the 'Abbey' series as 'Stars'. 5084 was further rebuilt in 1958 with 4-row superheater and double chimney and was withdrawn in 1962. (MLS Collection)

Rebuilt 'Abbey'
5091 *Cleeve Abbey*, converted in December 1938 and withdrawn in 1964. It is pictured here at Cardiff General station with the Down *Pembroke Coast Express* in June 1960. It has been fitted with a mechanical lubricator ahead of the steam pipe and a Hawksworth flush-sided tender. It can be recognised as a rebuilt 'Star/Abbey' only by the narrow rectangular inside cylinder casing as these engines retained their 'joggled' front end frames. It is a Landore engine, recognised as such immediately by the silver painted buffers. (GW Trust)

5081, built in 1939, named *Penrice Castle* and renamed *Lockheed Hudson* in 1941, on a Down express near Miskin between Cardiff and Bridgend at the end of the Second World War, c1947. It is still fitted with a metal plate cab window fixed during the war period to reduce the glare from the fire and the risk of being spotted and attacked by enemy planes. (J.G.Hubback/John Hodge Collection)

100 A1 *Lloyds,* rebuilt from 4009 *Shooting Star* in 1925 and renamed in 1936, was selected as one of five Castles converted to burn oil between 1946 and 1948 – the others were 5039, 5079, 5083 and 5091. 5091's fuel tanks were mounted on a 3,500 gallon tender but the rest retained a 4,000 gallon tender with the tanks as seen here. 100 A1 is at Old Oak Common where it was based in 1946. (W. Potter/MLS Collection)

The last of the '5013' series, 5097 *Sarum Castle,* built in July 1939. It is still with the original styled outside steam pipes and rectangular inside cylinder block. It is departing from Dawlish with the 9.5am Liverpool-Plymouth, the Shrewsbury/Newton Abbot double home engine and men turn. 5097 was equipped with a double chimney in July 1961 and withdrawn in 1963. (R.O. Tuck/Rail Archive Stephenson)

Wolverhampton Stafford Road's 5026 *Criccieth Castle* at Cardiff General station in the late 1950s. It is displaying here the chunkier outside steam pipes modified in the 1950s to minimise cracks and steam leakages. It also has the raised step over the inside cylinder block which appeared on Castles from the mid-1950s onwards. It too was fitted with 4-row superheater and double chimney in October 1959 and was withdrawn in November 1964 from Oxley LMR shed. (GW Trust Collection)

Canton's 5099 *Compton Castle* at Cardiff General with a Gloucester-Swansea stopping train, 31 March 1956. 5099 was the second of the Hawksworth 3-row superheater Castles built in May 1946, originally with a Collett 4,000 gallon tender but here seen with the Hawksworth flush-sided tender. 5099 would be transferred to Old Oak Common in November 1956 while the Canton shedmaster was on leave much to that gentleman's wrath, as he considered it Canton's best Castle. It was the first of the Hawksworth Castles to be withdrawn in February 1963 from Gloucester shed. (R.O. Tuck/Rail Archive Stephenson)

Laira's 4087 *Cardigan Castle* on Cardiff Canton shed, c1959. 4087 was built in 1925, but received a new front end frame in 1950, a 4-row superheated boiler in 1956 and a double chimney in February 1958. It was the fourth Castle to receive a double chimney after 7018, 4090 and 4093. It was withdrawn in October 1963. Sixty-six Castles were fitted with 4-row superheat boilers and double chimneys between 1956 and 1961. (John Hodge)

Chapter 2
THE LAST YEARS AT THE TOP – 1960–1962

The British Transport Commission's intention was to build 300 diesel locomotive prototypes, test them thoroughly and then mass produce the most efficient and successful as an interim step to electrification. The first were English Electric Type 4 (2,000 hp) diesel electrics mainly for the West Coast main line, though a few were spared for the East Coast and the Liverpool Street-Norwich services. The Brush Type 2 D55XX made early inroads in 1958 into the East Anglian 'B1s' and B17 'Sandringhams' in particular, but the Western Region Board, as independent as ever, persuaded the BTC to let them test the diesel hydraulic concept on the basis of the successful West German experience of the 1953 built V200 and the clear advantage they seemed to offer in the power/weight ratio (the 'Class 42 Warships' weighed 78 tons compared with the 'Peak' class 44's 138 tons). The Swindon management obtained the licence to build the West German Maybach/Mekydro design, the WR engineers adapting the V200 to the British loading gauge and D800 appeared in August 1958 and after initial trials worked its first Laira based diagram involving the Up *Cornish Riviera Express* in October 1958. D801 and 802 were built by the end of the year and D803-813 were built in 1959. The real influx came in 1960, with D814-829 with Maybach engines and the first ten class 43 MAN/Voith engines, D833-842. The five North British D600 Co-Co locomotives also came in 1958 but were unreliable from the start and made little impact on steam working, though they did take over (on paper at least) a couple of Laira and Old Oak King diagrams.

The initial intention had been to use the Maybach engines on the Birmingham route but when the GW line to Birmingham was made the prime route while the West Coast main line was electrified, both train frequency and loads were increased to the extent that it was thought the diesel hydraulics had insufficient power and 'Kings' were diagrammed instead. The WR operating management therefore decided to use the new diesels to blanket services west of Newton Abbot to avoid the costly double heading of steam power over the Devon and Cornwall banks. It didn't quite work out that way as the Class 42s in 1959 and 1960 were based at Bristol Bath Road and Laira new diesel depots, with the diesel engines working right through from London to Plymouth and Penzance via both the Berks & Hants and Bristol. At first the 'Warships' were limited to 90mph, then were allowed 100mph when the *Bristolian* was accelerated to 100 minutes. However, the WR Civil Engineer had concerns about rail damage (problems had been experienced in the USA) with the weight borne on just two small wheeled bogies and an 80mph limit was imposed in 1960 until bogie modifications were made and maximum speed raised to 90 in 1961. This all followed five years (1954-59) when alone on BR the Western Region's main lines had no overall maximum line speed limit (I think the working timetable quoted the words 'as high a speed as is necessary to maintain the timetable').

Another major change came in 1960. The BR train planners increasingly adopted the even

One of Laira's last two operational Castles, 4087 *Cardigan Castle*, still working in Cornwall despite the plan to fully dieselise Devon and Cornwall as priorities, with a Down express for Penzance at Par, June 1961. 4087 has 4-row superheater, double chimney and Davis & Metcalfe valveless mechanical lubricator. (J. Leeson/MLS Collection)

interval timetable system run successfully on the Swiss national railway (the 'Takt Fahrplan') for years. This rejected the occasional very fast 'flyer' that was superior to the rest of the service in favour of regular interval services that went every hour or two hours on the same clock face times, so that passengers could rely on knowing they had a train at 'x' minutes past the hour. The GW had some regularity – most West of England trains left Paddington at xx.30 and the South Wales trains had left on 55 minutes past the hour for years, but their timings down the line were erratic – the 10.55am *Pembroke Coast Express* times bore no relationship to the heavier, slower 11.55 or 1.55 and there was no 12.55. This new type of timetable was far more suited to the consistency of diesel loco performance and minimised the delay caused by extra stops with their greater acceleration. They therefore did not require the selection of specially prepared engines for the best trains that the WR had perfected for trains like the *Bristolian*, *Torbay Express* and the *Cambrian Coast Express* in both directions, where Old Oak Common, Bristol Bath Road and Newton Abbot would select a recently ex-works Castle to run the diagram for weeks at a stretch.

1959 was the last year in which steam remained on the fastest trains to Bristol or Exeter. By 1960, there was only one 'King' diagram left in each direction between Paddington and Plymouth – the 11.30am Down and the 12.5pm Plymouth Up, for which Laira had two 'Kings', 6002 and 6016. The 11.30am Down in particular was often hauled by an Old Oak Common 'King' or Castle or even a Laira Castle. The Laira 'Kings' migrated to Canton, Old Oak and Stafford Road and the Laira Castles gradually disappeared to other haunts until by 1962 just two were left, the double chimney 4087 *Cardigan Castle* and 7022 *Hereford Castle*, which acted as standbys for diesel failures. I remember seeing a photo in *Trains Illustrated* in January 1962 of 4087 in full cry on the Up *Mayflower* overtaking a 'Warship' on a Bristol train standing in Reading station. Of course, in summer the situation was different. Holiday traffic to Devon and Cornwall was still at its peak and a typical day was analysed by Richard Woodley in a book I commend called *The Day of the Holiday Express*. Most of the extra relief services would be steam hauled until the end of the 1961 summer service and some would still be steam in 1962.

Old Oak Common's travel-stained 5014 *Goodrich Castle* assists 6836 *Estevarney Grange* with a relief to the *Cornishman*, on Rattery Bank, the late May Bank Holiday weekend, 27 May 1961. (MLS Collection)

Newton Abbot's 5072 *Hurricane* emerges from Parson's Tunnel, Dawlish, with the Down *Cornishman*, 19 August 1961. (A.C. Gilbert/ MLS Collection)

5085 *Evesham* Abbey, brought out of store for the summer peak, brings a holiday makers' train from the West Midlands to Paignton beside the River Exe between Exminster and Starcross, c1961. (J. Davenport/MLS Collection)

Old Oak Common's 5032 *Usk Castle* on what appears to be the *Royal Duchy*, the 1.30pm Paddington-Penzance, passing through Tiverton Junction, 22 May 1961. It would seem to be a substitute for a failed 'Warship' diesel hydraulic which would have been diagrammed to this service by 1961. (MLS Collection)

4098 *Kidwelly Castle* arriving at and departing from Exeter St David's with the 9.10 Manchester-Plymouth, acting as a relief on Saturdays to the weekday 9.5 Liverpool double-home diagram, 19 August 1961.
(A.C. Gilbert/MLS Collection)

7022 *Hereford Castle*, the other one of Laira's last working pair of Castles, on a Down fitted-freight at Aller Junction, 26 May 1960. (MLS Collection)

Old Oak Common's 7010 *Avondale Castle* heads an Up West of England express through Southall station in the summer of 1961. (J. Davenport/MLS Collection)

As we move from 1959 to 1960, on a normal weekday we find the main Paddington-West of England and London-Bristol services to be hauled by 'Warships' of both Swindon and North British varieties. The South Wales trains are still steam but the 'Britannias' will soon be moving to the London Midland Region as they are replaced by redundant West Country 'Kings' and the increasing number of double chimney Castles. The Gloucester and Worcester/Hereford services are firmly in the hands of the Castles as are most of the Bristol-Shrewsbury North & West expresses. The augmented Birmingham/Wolverhampton service is mainly in the hands of the 'Kings' backed up by the fleet of Castles still remaining at Old Oak and Stafford Road. This is all reflected in the various Castle transfers that took place in 1960 and the allocation at the beginning of the Summer 1961 timetable was (increase or decrease from 1959 shown in brackets, double chimney Castles shown*):

Old Oak
  Common: 4075, 4082 (ex-7013), 4089, 4096, 5001*, 5008*, 5011, 5014, 5015, 5032*, 5034*, 5036*, 5040, 5042, 5043*, 5056*, 5057*, 5060*, 5065, 5066*, 5082, 5084*, 5087, 5090, 5093, 7001*, 7009, 7010*, 7017, 7020*, 7024*, 7030*, 7032*, 7033*, 7036*, 7037 (36, +2)
Reading: 4086, 5018, 5067, 5076 (4, -1)
Oxford: 4092, 5012, 5025, 5033*, 7008* (5, +1)
Bristol
  Bath Road: 7019* (1, -17)
Bristol
  St Philip's
  Marsh: 4077, 4079, 5049*, 5050, 5052, 5085, 5094*, 7014*, 7018*, 7034* (+10)
Swindon: 4074*, 4088*, 5000, 5002, 5023, 5035, 5064*, 5068*, 7031 (9, +1)
Newton
  Abbot: 4037, 4098, 5003, 5024, 5055, 7029* (6, -7)
Taunton: 5073*, 5096 (+2)
Exeter: 4083, 5020, 5075 (3, +2)
Laira: 4087*, 4095, 5029, 5053, 5058, 5069*, 5098*, 7022* (8, -2)
Penzance: (-2)
Stafford
  Road: 5019*, 5022*, 5026*, 5031*, 5045, 5046, 5047, 5063, 5072, 5088*, 5089, 7012, 7026 (13, -1)
Banbury: (-3)
Shrewsbury: 5038, 5059, 5070, 5095*, 7015*, 7025 (6, +2)
Worcester: 4085, 7002*, 7004*, 7005, 7006*, 7007*, 7011, 7013* (ex-4082), 7023*, 7027 (10, -1)
Gloucester: 5007, 5017, 5071*, 7000, 7003*, 7035* (6, +2)
Cardiff
  Canton: 4080*, 5021, 5061*, 5081, 5092, 5097*, 5099 (7, no change)
Neath: 4093*, 4099, 5013, 5037, 5041, 5044, 5048, 5051, 5062, 5074, 5078 (+11)
Landore: 4076, 4090*, 4094, 5004, 5030, 5080, 7021, 7028 (8, -10)
Llanelli: 4078, 5006, 5016*, 5077, 5091, 7016 (+6)
Carmarthen: 4081, 5027*, 5039, 5054 (4, -3)

Total (155, -9)

With the closure of Bath Road shed, almost all of its Castles had moved to St Philip's Marsh and although Laira's diesel depot had been completed alongside the steam shed, a fair number were still there until the end of the 1961 summer timetable, though several in store. The Newton Abbot fleet had shrunk, although both Exeter and Taunton had gained a few. Landore diesel depot was being constructed and most of its fleet relocated to Neath and Llanelli, the lower mileage engines to Neath. With main line steam services planned to last longest on the Worcester trains, that depot exchanged many of its Castles for the post-war 70XX series, many already with double chimneys. Twenty-one more Castles had received a double chimney by July 1961, leaving just 5074, 5078, 5092, 7021 and 7028 to be converted in the autumn, the last being 5078 before a decision was made by the new Regional management of Stanley Raymond and Lance Ibbotson that no more money would be spent on main line WR steam power as full dieselisation of express services would be completed by 1963.

Double chimneys were fitted to the following Castles in 1960 and 1961:

1960
January: 7035
February: 5016, 5034
May: 7006
June: 5094, 5097, 7003
September: 7001, 7032
October: 5033, 5056, 7010
December: 5036

1961
February: 5019, 7020
March: 5008, 7007
April: 5027
June: 5001, 7002
July: 5060
September: 5074
October: 5092, 7028
November: 7021
December: 5078

Sixty-six Castles were finally fitted with double chimneys at a time when it was envisaged that steam on trains like the Paddington-Gloucesters and Worcesters would remain until the late 1960s. This included ten of the 1924-1927 built engines, 4073-5012 (including 7013, ex-4082). All the new double-chimney Castles of this series apart from 4074 had new front end frame sections and all had a 4-row superheat boiler. Many people considered these old 'renewed' engines some of the strongest. However, 4097 *Kenilworth Castle* fell an early victim to severe frame cracking (it had been the first to have a new front end before 1950) and was surprisingly withdrawn in May 1960, less than two years after receiving the double chimney. Its boiler will have been recovered and reused. Twelve of the last engines to be fitted with a double chimney also had less than two years in that condition – 5001, 5008, 5019, 5027, 5033, 5036, 5060, 5078, 5092, 5097, 7007 and 7021.

It is time now to have a look at Castle performance in the

Old Oak Common's 5036 *Lyonshall Castle* rebuilt with 4-row superheater and double chimney in December 1960 and seen here on the 2.55pm Paddington-Swansea at Newport High Street, 9 May 1961. 5036 was withdrawn as redundant at the cull of Castles in September 1962. (John Hodge)

Landore's 4097 *Kenilworth Castle* at Cardiff Canton shed, c1959. It was rebuilt with double chimney in June 1958 and was the first double chimney Castle to be withdrawn in December 1960. It is believed to have been the first of the 40XX Castles (with 4087) rebuilt with a new front end frame section as early as 1950 and it is probable that the reason for early withdrawal at the first planned overhaul after the rebuilding was due to cracked frames. (John Hodge)

final years. Did it deteriorate or did it hold up to the last as the performance of the 'Kings' on the Wolverhampton road did? One problem is that the named trains on which the performance of Castles had been often recorded were either dieselised or left the timetable after the new even interval services were introduced. Opportunities for recording exceptional performance were limited and restricted to the South Wales route until dieselisation in 1962, then the North and West north of Bristol and the Gloucester and Worcester routes. The Western Region had 36 scheduled daily runs timed at over 60mph start-to-stop in the winter 1961 timetable, more than any other Region, though 33 of these were diesel hauled runs to Bristol or Taunton/Exeter and the diesel 'Blue Pullman' services to Newport and Leamington. The three steam-hauled fast services were the Down *Pembroke Coast Express* to Newport at 61.1mph, the Up *Cheltenham Spa Express* from Kemble to Paddington at exactly 60mph and the 5.30pm Oxford-Paddington at 63.4mph. I therefore persuaded a college friend, Alastair Wood, to join me on a trip to Newport and back on the *Pembroke Coast Express* on a Saturday when the turn would be harder with additional load. What we did not know was that the WR authorities had a bad practice of retiming the fastest trains if over the tonnage limit but not advertising the change to the public. We had Neath's 5037 *Monmouth Castle,* an engine with a good reputation, with a substantial load of 11 coaches, 420 tons gross, and waited to see how it would cope with the sharp schedule. It never tried. We proceeded to dawdle all the way to Swindon taking 91 minutes with a short signal stop at Reading without exceeding 65mph before Reading, 59mph between Reading and Didcot, and 56mph between Didcot and Swindon, with remarks in my notebook that 5037 was blowing off steam furiously the whole way, drowning out the noise of any exhaust. We fell to 50mph at Badminton with 5037 still blowing off on the whole ascent (goodness knows what the fireman thought of his wasted steam) and the driver finally woke up with 85mph in the middle of the Severn Tunnel to prove that there was nothing inhibiting him in the riding of the engine. We crawled into Newport in 153 minutes 25 seconds, 22½ minutes late by the public book. My colleague tackled the driver at Newport, asking him what was wrong with the engine and why we were late. 'There's nothing wrong with the engine,' the driver snapped back. 'We're on time!'

The Up *Pembroke Coast Express* was allowed 144 minutes for the 133 miles, inexplicably slower than the harder Down run, and after our morning experience, Alastair was pessimistic about our return, which ran into Newport nearly 10 minutes late behind 1924 built 4081 *Warwick Castle* looking ex-works. It was not as bad as we feared.

Newport-Paddington, 28.10.1961
4081 *Warwick Castle* Llanelli
Pembroke Coast Express
10 chs, 347/375 tons

Miles	Location	Times	Speeds	Gradients
0	Newport High ST	00.00	9 ¼ L	
	Llanwern	06.16	54/62	
7.4	Magor	11.92	sigs 12*/ 10*	
9.7	Severn Tunnel Jcn	16.56/18.04	Special stop	
10.9	Severn Tunnel West	20.42	52/68	1/90 F
15.2	Severn Tunnel East	24.49	74/51	1/100 R
16.7	Pilning	26.41	45/47	1/100 R
20.3	Patchway	32.12	35	1/68 R, 1/90 R
23.5	Winterbourne	36.22	55	1/300 R
28.8	Chipping Sodbury	42.15	58/55	1/300 R
33.4	Badminton	47.22	55	1/300 R
39.1	Hullavington	52.43	76 ½	1/300 F
43.6	Little Somerford	58.10	pws 10* (½ mile)	
50.3	Wootton Bassett	68.16	54	

Newport-Paddington, 28.10.1961
4081 *Warwick Castle* Llanelli
Pembroke Coast Express
10 chs, 347/375 tons

Miles	Location	Times	Speeds	Gradients
56.1	Swindon	74.20	61	
61.9	Shrivenham	79.35	72	
66.9	Uffington	83.42	77	
69.5	Challow	85.50	79	
73	Wantage Road	88.30	78	
76.8	Steventon	91.30	77	
80.2	Didcot	94.04	78	
84.9	Cholsey	97.47	79	
88.6	Goring	100.41	77/74	
91.8	Pangbourne	103.22	62 eased	
97.5	Reading	111.22	pws 20*	
102.5	Twyford	118.17	64	
109.3	Maidenhead	124.49	68	
115.1	Slough	130.08	69	
124.5	Southall	138.40	67	
127.9	Ealing Broadway	141.39	70	
132.1	Westbourne Park	146.34	sigs	
133.3	Paddington	150.41	(133 net)	15 L

The reason for the special stop at Severn Tunnel Junction is unknown. The climb out of the tunnel and the running from Swindon to Reading was excellent, though the last stretch from Reading was a little disappointing. 4081 was working hard but was possibly a little stiff being so recently ex-works. Alastair and I thought we must have caught 5037's driver on a bad day and tried our luck again on a couple of other Saturdays but fared no better. We had good engines on both occasions, Neath's double chimney Castles 5078 *Beaufort* and 7028 *Cadbury Castle,* but both lost 15 minutes on the public timetable and both times the engines gave signs of having a lot of unused steam wasted through the safety valve. 5078 did have 12 coaches, 440 tons gross, but we maintained 60-65mph all the way from a p-way slack at Taplow to Wootton Bassett without any apparent effort (from later observation, I guess 15 per cent cut-off and first valve of the regulator only). We reached Newport in 146½ minutes (142 net) and I feel could so easily have made Newport in 135 minutes or less. A few months after our attempts, another recorder, Mr B. Nathan, was more fortunate and got Neath's 4099 *Kilgerran Castle* with a 10 coach train and reached Newport in exactly 136 minutes including two p-way slowings (131 net, the public scheduled time), and Mr D. Twibell did even better with double chimneyed 5016 *Montgomery Castle* and a nine coach load which was over a couple of minutes up on 4099 at Severn Tunnel Junction but got badly checked at Maindee East Junction and took 142 minutes (129 net).

		Paddington-Newport (*Pembroke Coast Express*)				
		5016 *Montgomery Castle*		4099 *Kilgerran Castle*		
		Neath		Neath		
		9 chs, 330 tons		10 chs, 343/370 tons		
		30.6.1961		7.10.1961		
Miles	Location	Times	Speeds	Times	Speeds	Gradients
0	Paddington	00.00		00.00		
1.3	Westbourne Park	03.06		03.34		
2.7	Old Oak Common W	-	pws 24*	-	pws 15*	
5.7	Ealing Broadway	10.18	52	11.34	51	
9.1	Southall	13.43	62	15.24	55	
13.2	West Drayton	17.29	69	19.38	62	
18.5	Slough	21.51	73	24.25	67	
24.2	Maidenhead	26.36	72	29.36	67	
31	Twyford	32.12	71	35.34	pws 33*	
36	Reading	37.20	sigs 49*	41.28	sigs 25*	
41.5	Pangbourne	42.47	64	48.43	62	
44.7	Goring	45.40	67/pws 15*	51.45	65	
48.5	Cholsey	51.17	30	55.05	67	
53.1	Didcot	56.35	61	59.12	68	
60.4	Wantage Road	63.19	65	65.42	65	
63.8	Challow	66.24	66	68.48	65	
66.5	Uffington	68.52	63	71.17	64	
77.3	Swindon	78.35	69	81.08	67	
83	Wootton Bassett	83.39	54*	86.38	53*	
89.7	Little Somerford	89.57	72	92.53	73	1/300 F
94.2	Hullavington	94.02	60/56	97.01	56	1/300 R
100	Badminton	99.55	59	102.57	64	1/300 R, L
104.5	Chipping Sodbury	104.05	66	107.13	66	1/300 F
108.4	Coalpit Heath	107.30	70	110.13	75	1/300 F
109.7	Winterbourne	108.43	65	111.37	71	
	Filton Junction	-	43*	-	48*	
113	Patchway	112.23	74	115.01	75	1/90 F
116.7	Pilning	115.50	68	118.24	75	1/100 F
118	Severn Tunnel East	-	74/80	-	80/81	1/100 F

		Paddington-Newport (*Pembroke Coast Express*)					
		5016 *Montgomery Castle*		4099 *Kilgerran Castle*			
		Neath		Neath			
		9 chs, 330 tons		10 chs, 343/370 tons			
		30.6.1961		7.10.1961			
Miles	Location	Times	Speeds	Times	Speeds	Gradients
122.4	Severn Tunnel West	-		123.32	46	1/90 R
123.5	Severn Tunnel Jcn	122.30	34	124.48	40	
125.9	Magor	125.58	61	127.55	63	
132.6	Maindee East Jcn	-	sigs stops	134.47		
133.3	Newport High St	142.09 (129 net) 11L		136.02 (131 net)	5L	

Landore's 5039 *Rhuddlan Castle* (just transferred from Carmarthen and not yet 'decorated' with silver buffers) heads the 10.55am Paddington Saturday *Pembroke Coast Express* at Pengam, with four strengthening ex-LMS/GW coaches at the front of the nine regular chocolate and cream set. It will have been retimed in the working timetable as was the WR's practice at the time. (R.O. Tuck/Rail Archive Stephenson)

Landore's double chimneyed 5016 *Montgomery Castle* at Cardiff General with the Down *Pembroke Coast Express*, 2 March 1961. Note the post-1960 revised train description – the numeric code was replaced by a letter indicating the terminating Division. 'F' was South Wales, 'A' London, 'C' Plymouth Division, 'B' Bristol, 'O' to the Southern Region, 'M' to LMR and 'V' to WR from another Region. (John Hodge)

The Last Years at the Top – 1960–1962 • 31

Landore's 4099 *Kilgerran Castle* on arrival at Cardiff General with the Down *Pembroke Coast Express* in February 1959. 4099 was moved to Llanelli after the closure of Landore and Roy White, shedmaster at Llanelli in the early 1960s, believed it to be his best Castle. (John Hodge)

Double chimney 5097 *Sarum Castle* of Canton heads the Up *Red Dragon* through Reading station, 5 September 1961. It will return to South Wales on the 3.55pm Paddington *Capitals United Express*. (MLS Collection)

Old Oak Common's 7036 *Taunton Castle* arrives at Shrewsbury with the Down *Cambrian Coast Express* with its weekday fast return locomotive working from Paddington, c1961. As the train would be strengthened to twelve or thirteen coaches on a Saturday it was normal to diagram a 'King' on those days. Note, however, that on this occasion at least one strengthening vehicle (a Gresley coach) has been added to the regular chocolate and cream set. (John Hodge)

5062 *Earl of Shaftesbury* winds into Newport High Street with the Up *South Wales Pullman* on a glorious May evening, 9 May 1961. (John Hodge)

On another occasion with the *Pembroke Coast Express* in January 1961, 5077 *Fairey Battle* with 10 coaches, 365 tons, was going reasonably well to Reading, passed at 71mph, but had a dreadful time of it to Didcot. After a severe p-way restriction at Cholsey, it had a number of signal stops between Moreton Cutting and Steventon and took 39 minutes to cover the 24.4 miles from Reading to Wantage Road. No sooner had 5077 accelerated to 66 at Uffington when it suffered another p-way slack to 15mph at Shrivenham, taking nearly 100 minutes to pass Swindon. 5077's driver then set about trying to recoup some of the lost time and made an excellent climb to Badminton with 77 at Little Somerford, 66 at Hullavington and a minimum of 64 at Badminton. It then passed through the Severn Tunnel in four minutes exactly touching 80 in the tunnel after signal checks to 50 at Pilning and was still doing 52mph on the 1 in 90 out of the tunnel, and had to brake for the junction. More signal checks at Maindee Junction and arrival in Newport was still 28 minutes late by the public book (13 minutes by the revised timing). Net time was around 130 minutes.

Landore's 4076 *Carmarthen Castle* accelerates the 8.55am Paddington-Pembroke Dock out of Cardiff, 27 February 1960. This diagram involved the Up *Pembroke Coast Express* of the previous evening. (John Hodge)

Old Oak Common's 5056 *Earl of Powis* with an Up South Wales express at Wantage Road, 5 June 1960. 5056 would be rebuilt with a double chimney in November 1960.
(MLS Collection)

Canton's double-chimneyed 4080 *Powderham Castle* enters Cardiff General station with the *Capitals United Express*, 3.55pm from Paddington, c1961. It was the regular engine on the Up *Red Dragon*/3.55pm return at this time.
(John Hodge)

Landore's 5013 *Abergavenny Castle* draws out of Cardiff General with the 8.55am to Pembroke Dock, 16 March 1961. (John Hodge)

5093 *Upton Castle* of Old Oak Common at Paddington on the 11.55am to South Wales, 28 June 1961. (MLS Collection)

Britannia substitute, Canton's 5021 *Whittington Castle*, stands at Cardiff General with the 12noon Cardiff-Paddington, 9 March 1961. It will return on the 5.55pm Paddington *Red Dragon*. (John Hodge)

An old time Old Oak Castle, 5044 *Earl of Dunraven*, transferred to Canton in July 1960, leaving Cardiff General with the 11.50am Swansea-Manchester which it took over at Cardiff. (John Hodge)

The Last Years at the Top – 1960–1962 • 37

Old Oak Common's 4078 *Pembroke Castle* entering Newport High Street station with the 11.55am Paddington-Milford Haven, 5 March 1961. (John Hodge)

Old Oak Common's 5008 *Raglan Castle* entering Newport High Street with the 11.48am Cardiff-Paddington (relief to the 10.30am Swansea), 24 September 1960. 5008 would receive a double chimney in March 1961 and be withdrawn in September 1962. (John Hodge)

Carmarthen's 7016 *Chester Castle* at Llanelli with the Up *Pembroke Dock Express* which it will take as far as Swansea High Street, 26 May 1961. (MLS Collection)

I can only find one log of the Up 60mph post-war *Cheltenham Spa Express* in the Rail Performance Society archives, a January 1961 run with the shed's pet engine, 5017 *The Gloucestershire Regiment 28th 61st* and 10 coaches, 355/380 tons. The train left Stroud seven minutes late and was assisted to Sapperton summit by 'Large Prairie' tank 6137, with 44mph on the 1 in 70 at Chalford and 32 at the top of the 1 in 60. 5017, now on its own, left Kemble six minutes late and ran at a steady 75mph between two p-way slacks to 40mph at Uffington and Cholsey. The train roared through Reading at 72½mph and reached 77½ before Maidenhead and was still doing 75 at Slough when the rest of the run was ruined by signal checks – to walking pace before Southall and a complete stand at Subway Junction. The Paddington arrival was 11 minutes late, although the net time for the 91 miles from Kemble to Paddington was no more than 85 minutes and without the checks the late start from Kemble would have been recovered. I had one run myself on the Down 5pm Paddington *Cheltenham Spa Express* with Gloucester's double chimney 7035 *Ogmore Castle*. With a load of 9 coaches, 311/330 tons we sustained 70mph from West Drayton to Maidenhead and then suffered signal checks and a p-way slowing to 15mph from Twyford through Sonning to Reading. Speed then rose to 70 by Goring and 76 by Challow, falling slightly to 74 mph at Uffington. Swindon was passed in 83 minutes 51 seconds and after another p-way slack to 25mph at Purton, we took 102½ minutes to Kemble, net time 91 minutes for the 91 miles.

Gloucester's 4085 *Berkeley Castle* passes Ranelagh Bridge with the 2.15pm Paddington-Cheltenham. 4085 worked the Up *Cheltenham Spa Express* to Paddington earlier in the day, May 1959. (R.C. Riley/MLS Collection)

Gloucester's 7034 *Ince Castle* passing West Ealing with the Up *Cheltenham Spa Express*, 19 March 1962. 6100 is in the background. (Charles Gordon-Stuart/GW Trust)

The third 60mph booking was the 5.30pm Oxford-Paddington, booked for an Old Oak Castle on weekdays with a featherweight load of six coaches around 195/210 tons. This could be quite exciting as pre-war 'Cheltenham Flyer' style running was necessary from Didcot to Paddington to keep time. During management training at Oxford in March 1962, I found the most interesting way of returning to my lodgings in Reading was via the non-stop 5.30pm to London and the 7.5pm Paddington-Cheltenham back to Reading! I had a London Division free pass making this possible. I have eleven recorded runs on this train with the following overall times recorded for the 63.4 miles:

Date	Loco	Load (tons)	Act (mins)	Net (mins)	
1/3/62	7014	161/170	59.58	58 net	
2/3/62	5060	161/175	57.40	57¾	
7/3/62	5066	189/205	71.01	67	(short of steam between Maidenhead & S'hall)
8/3/62	7021	189/205	61.50	56½	
9/3/62	5076	189/210	61.10	57½	
19/3/62	7018	189/200	62.18	62¼	
20/3/62	5034	189/200	64.15	61	
21/3/62	5036	189/210	60.36	60½	
23/3/62	5015	189/210	58.18	58¼	
18/4/62	5001	188/200	53.09	52½	
16/5/62	7030	188/200	56.19	53½	

Old Oak Common's 4089 *Donnington Castle* at Oxford with the 5.30pm 60-minute non-stop express for Paddington, 10 August 1962. This is the standard set for this train – I travelled many times in that Hawksworth BSK coach immediately behind the engine. (Charles Gordon-Stuart/GW Trust)

The loss of time with 5066 on 7 March was caused by a clinkering of the fire – the engine had, according to the driver, been left unattended for too long on Oxford shed and after a blazing start with 76mph by Culham and 78 by Goring, steam pressure dropped rapidly and speed fell to 55-60 between Reading and Slough. Hard work by the fireman improved things by West Drayton and the driver opened up 5066 again to 72 by Acton before signal checks in from Old Oak added another four minutes over schedule to the run. I will cover the two very fast runs in the chapter on my personal experiences (pages 60 and 72) as I was on the footplate of 7030 in May, but I will show in the table below a couple of my fastest March runs plus a log of 5056 in August that year timed by Mr Twibell.

Oxford-Paddington, 5.30pm Oxford

		7021 *Haverfordwest Castle* 188/205 tons 8.3.1962		5076 *Gladiator* 188/210 tons 9.3.1962		5056 *Earl of Powis* 187/205 tons 21.8.1962	
Miles	Location	Times	Speeds	Times	Speeds	Times	Speeds
0	Oxford	00.00		00.00		00.00	
5.1	Radley	06.50	72	07.13	69	06.45	68
7.3	Culham	08.50	74/77	09.14	73/77	08.42	72/75
10.6	Didcot East Jcn	11.57	38*	12.23	40*	11.49	44*
14.9	Cholsey	16.28	78	17.05	76	16.07	75/80
21.9	Pangbourne	22.03	82/85	22.34	84/86	21.19	84/88
27.4	Reading	26.18	82	26.41	80	25.17	83
32.4	Twyford	30.13	87/75*	32.40	sigs 2*/72	28.59	80/75
39.2	Maidenhead	35.36	86	39.09	82	33.57	90
44.9	Slough	39.56	87/84	43.33	86	37.55	90
	Iver	42.43	83/sigs 23*	46.25	82	40.29/40.47	sig stand
50.2	West Drayton	44.45	51	47.38	81	43.57	57
54.3	Southall	49.08	79	51.00	82/85	47.29	80
57.7	Ealing	51.50	84/sigs 32*	53.37	86	50.01	80/75
62.1	Westbourne Park	56.52/57.20 sig stand		57.47		54.16	
63.4	Paddington	61.50 (56½ net) 1¾ L		61.10 (57½ net) 1¼ L		57.37 (55 net) 2½ E	

Whilst training at Oxford I also used the Worcester line services regularly between Reading and Oxford. Most services were in the hands of Worcester based Castles, the exception being an Old Oak turn on the 9.15am Paddington and the 12.05pm Hereford return, the Old Oak engine turning round at Worcester. The schedules demanded competent rather than exceptional performance and, provided the signalmen allowed it, the Castles could keep time easily enough maintaining speeds at around 68-70mph between Paddington and Oxford. A typical Down performance was that of Old Oak's double chimney 5066 *Sir Felix Pole* on 5 March 1962, which I caught through from Paddington. With 8 coaches, 278/300 tons we would have made the 38 minutes schedule to Reading (36 miles) but for a signal check outside Reading station. We actually took 39 minutes 5 seconds. After a fairly easy start

(13½ minutes to Southall), we passed 70mph by West Drayton and ran at 72-74mph between Slough and Twyford. Leaving Reading a couple of minutes late, 5066 ran the 27.4 miles to Oxford unchecked in 31 minutes 49 seconds (schedule 37 minutes), arriving at Oxford 3 minutes early. Maximum speed was 72mph just before the slowing for Didcot East Junction.

Another interesting occasion was catching Worcester's 5099 *Compton Castle* (an erstwhile favourite of both Canton and Old Oak) on the 11.15am from Paddington on 26 February 1962 when we ran through a heavy snow storm between Paddington and Reading. The load was similar (295 tons gross) and the driver did not press 5099 beyond 67mph en route to Reading as conditions must have been unpleasant and vision limited. 5099 therefore exceeded the 38 minute schedule by a couple of minutes, but the driver presumably was aware that the 37 minute schedule to Oxford included 4 minutes 'recovery time'. However, the driver would not have anticipated a dead stand at Scours Lane leaving Reading, nor a signal check almost to a stand at Didcot East Junction, so we took the full scheduled time and arrived 3 minutes late. In the heavy snow this was at least acceptable for passengers and showed the usefulness of the GW ATC system in conditions of poor visibility.

Performance in the Up direction tended to be brisker. 7032 *Denbigh Castle* of Old Oak on the 12.5pm Hereford with 9 coaches, 312/330 tons cut the 34 minute schedule to Reading by 3½ minutes with 75mph from Goring to Tilehurst, then dashed for home with 79 at Maidenhead, eased to 68 after Slough, and then another spurt to 77 at Ealing Broadway arriving 3½ minutes early at Paddington in 37 minutes 16 seconds. 5067 *St

4074 *Caldicot* Castle on arrival at Reading with the 4.35pm Paddington-Swindon semi-fast train, October 1961. This was my regular commuting train back to Reading after my initial station training at Maidenhead. The changed shape of the inside cylinder block whilst retaining the original 'joggled' frame is very clear in this photo. 4074 received the double chimney in April 1959 and was withdrawn in May 1963. (David Maidment)

Fagans Castle of Reading was borrowed by Worcester for the 10.5am Hereford on 22 March and with eight coaches (295 tons) made very similar times, 31 minutes to Reading and 38 minutes to Paddington, maximum speeds 73 at Culham, 74 at Pangbourne, 74 at Maidenhead and a final fling at 79 through Ealing.

Oxford had four Castles and used one regularly on a morning commuter service, the 7.30am Oxford-Paddington, the engine slumbering all day at Old Oak Common before returning on the heavy 6.5pm Paddington-Oxford in the evening. I used the 7.30 Oxford for my travels between Reading and Slough in November and December 1961 whilst training at Slough Goods depot and performance was very reliable and consistent. Our usual engine was 5025 *Chirk Castle* behind which I recorded ten runs. Schedule for the 17.5 miles was 21 minutes start-to-stop and 5025 made it on every occasion except on 6 December when it suffered a p-way slowing to 13mph at Twyford and had a severe signal check outside Slough. Speed reached was usually between 68 (minimum) and 74 (maximum) around Maidenhead. The other Oxford Castles to appear were 5012 *Berry Pomeroy Castle* between 28 and 30 November which actually achieved my fastest run – 19¼ minutes with 74 mph between Maidenhead and Burnham, and one solitary run with 5033 *Broughton Castle* which had attained 71 by Ruscombe and then had a series of signal checks taking 24 minutes.

During the time I trained at Reading station in January and February 1962, I became very acquainted with the 'Conti' ('Continental' as it called at Dover), the nickname of the through Birkenhead-Margate/Hastings/Brighton service, which was hauled by a Reading Castle between Reading and Chester and return, handing over to a Southern engine at Reading (a 'Schools' in 1961/2). The southbound service was always overtime in the station while a member of staff pulled the brake cords on the coaches as the Southern engine could never overcome the 25in vacuum of the GW engine. This service remained steam at least until the end of 1962 as did Bournemouth-Wolverhampton/Birkenhead trains that used the Reading West curve avoiding Reading General station. Oxford Castles shared those turns with Reading engines, changing from the SR engine at Basingstoke. Summer Saturday 1963 trains from the West Midlands to the South Coast would remain steam.

5093 *Upton Castle* slips under the West London line at North Pole Junction/Old Oak Common East with a morning commuter train from Didcot and Reading, c1961. (MLS Collection)

A shed scene at Wolverhampton Stafford Road with Old Oak Common's 4082 *Windsor Castle* (ex-7013), Reading's 4074 *Caldicot Castle* on the left, Stafford Road's own 5072 *Hurricane*, a 'King', a 'Grange', a '51XX' prairie tank and another Castle, 21 October 1962. (MLS Collection)

Reading's 4092 *Dunraven Castle* at Southam Road on the 'Conti', the Birkenhead-Margate/Hastings/Brighton cross-country train, 20 June 1960. (MLS Collection)

Old Oak Common's 4096 *Highclere Castle* on the 'Conti' Birkenhead-Kent & Sussex Coast express at Fenny Compton, 20 June 1960. (MLS Collection)

Oxford's 7008 *Swansea Castle* passing West Bromwich with a Birkenhead-Bournemouth express of Southern stock, 15 September 1960. (MLS Collection)

4074 *Caldicot* Castle with the Up 'Conti' at Fosse Road, 25 August 1962. The 1923 built engine has been rebuilt with 4-row superheater and double chimney but has retained its original 'joggled' frame. (MLS Collection)

Oxford's 5068 *Beverston Castle,* with double chimney and extended smokebox, near Fosse Road with the 9.30am Bournemouth-Wolverhampton, 25 August 1962. A,C. (A.C.Gilbert/MLS Collection)

Rebuilt 'Star'
5085 *Evesham Abbey*, based at Reading for just three months in the summer of 1962, near the summit of Hatton bank, with the 9.25am Margate 'Conti', 25 August 1962. (A.C.Gilbert/MLS Collection)

We now come to the Worcester and Hereford expresses which were nearly always hauled by Castles with an occasional intervention by one of Worcester's 'Modified Halls'. Cecil J. Allen reported a very competent run with Worcester's single chimney 7005 in *Trains Illustrated* in July 1961:

Evesham-Paddington

7005 *Sir Edward Elgar* Worcester

The Cathedrals Express

9 chs, 310/330 tons

Miles	Location	Times	Speeds		Gradients
0	Evesham	00.00		7 L	
2.35	Littleton	04.13			
4.9	Honeybourne	07.03	57 / pws 32*		1/126 R
9.75	Chipping Campden	15.22	34½ / 33		1/100 R
11.75	Blockley	17.38	61½ /50		1/145 F, 1/151 R
14.9	Moreton-in-Marsh	21.29		6½ L	

Evesham-Paddington
7005 *Sir Edward Elgar* **Worcester**
The Cathedrals Express
9 chs, 310/330 tons

Miles	Location	Times	Speeds		Gradients
0		00.00		5½ L	
4.4	Adlestrop	05.57	64½		L
7.1	Kingham	08.22	69	4 L	1/402 F
10.1	Shipton	10.54	72		1/388 F
11.4	Ascott	11.58	75		L
15.1	Charlbury	14.58	80½ /77		1/315 F, L
21.3	Handborough	19.42	82		1/250 F
24.5	Yarnton	22.09	pws 30*	2 L	
25.4	Wolvercote Jcn	24.23	sigs 20*		
<u>28.3</u>	<u>Oxford</u>	<u>28.26</u>		<u>1L</u>	
0		00.00		T	
5.1	Radley	07.06	61½		
7.3	Culham	09.17	66		
10.5	Didcot East Jcn	12.25	40*	1 E	
15	Cholsey	17.56	64		
18.7	Goring	21.20	68		
21.9	Pangbourne	24.08	70 ½		
24.75	Tilehurst	26.29	75		
27.4	Reading	29.04		3½ E	
	Sonning	-	pws 22*		
32.4	Twyford	37.03	65	1 L	
39.2	Maidenhead	43.01	75	1 L	
45	Slough	47.41	76½	¾ L	
50.2	West Drayton	52.46	sigs 47*		
54.3	Southall	56.38	66	1½ L	
57.7	Ealing Broadway	59.30	72½		
60.2	Old Oak Common W	61.41	sigs 45*		
62.2	Westbourne Park	64.53	sigs 28*	1¼ E	
	Royal Oak	-	sig stop		
<u>63.4</u>	<u>Paddington</u>	<u>70.32</u>	(net 62)	<u>½ L</u>	

7006 *Lydford* Castle was reallocated to Worcester on 7 July 1960 and a few days later it is seen passing Goring with an Up Hereford & Worcester express, July 1960. It has not yet been polished up to the standard of other Worcester based Castles.
(J. Davenport/MLS Collection)

7013 *Bristol* Castle (ex-4082) in more typical external condition at the head of the 5.15pm Paddington-Hereford *Cathedrals Express* at Oxford, 1962.
(MLS Collection)

7031 *Cromwell's* Castle on the Up *Cathedrals Express* near Evesham, May 1962. (Derek Penney)

7007 *Great* Western at speed at Iver with the Down *Cathedrals Express*, 25 August 1962. (Peter Fry/GW Trust)

Chapter 3
PERSONAL RECOLLECTIONS – 1960–1963

My summer vacation in 1959 was spent at Munich University where Bavarian 4-cylinder compound pacifics replaced the Castles in my off-campus time. I had spent a week experiencing some of the engines and men I'd got to know at Old Oak Common during a week of riding up and down to Reading as described in the previous book and likewise in 1960 to celebrate finishing my Finals, also recounted in my previous Castle book. The 1960 performance and punctuality was improved over the 1959 experiences – more of the services were hauled by double-chimney Castles by then, several recently ex-works. The rebuilding of Hayes bridge which had hampered runs in 1959 had also been concluded.

I joined British Railways in August 1960 and worked as a clerk in the WR London Division's Passenger Train Office at Paddington where for several months I had access to Guards' journals before they were sent to the Swindon Statistical Office and if time was slack, I would interest myself in looking to see what locomotives were working what trains and whether the guard had booked time gained or lost against the engine. I would spend lunchtime wandering round the station, my break coinciding with the 1pm arrival of the *Red Dragon* from Cardiff, usually early as the schedule was so slack. It was always a Castle – double-chimney 4080 *Powderham Castle* is the one I remember best because for days it was arriving exceptionally early. I think it was during this time that for a brief period 92220 *Evening Star* had its famous week on the train, though I do not remember ever seeing any of the Canton Britannias on it. I think they had departed for the LMR by then.

With my newly acquired 'privilege' (25 per cent rate) rail tickets earned after a month's

Canton's double-chimneyed 4080 *Powderham Castle* on the 7.30am Carmarthen-Paddington *Red Dragon* passes Hawksworth pannier tank 8470 hauling a local freight, 20 August 1960. 4080 was a regular on this train at that time, being seen by the author arriving at Paddington very early day after day. (R.O. Tuck/Rail Archive Stephenson)

employment, I could spend the evening after work some days treating myself to a steam run to Swindon, Banbury, Rugby or Peterborough. It was usually the latter two as I'd had so many opportunities already to travel behind 'Kings' and Castles, though I remember several Saturday trips to South Wales as described in the last chapter and in the summer seeking out 47XX 2-8-0s on the 12.5pm or 1.25pm Paddington to Kingswear and getting Castles or what was known as a Summer Saturday Old Oak Castle (a 'Modified Hall'). I had 5008, 5092 and 7037 on three attempts to Taunton or Exeter but although I timed the trains, performance was inhibited by general congestion on the Berks & Hants. 5008 *Raglan Castle* was one of Old Oak's best in the summer of 1961, one of the last to get a double chimney, and we set off with our heavy 485 ton load in style on the 12.5pm to Plymouth. 5008 had reduced our 10 minute late start to 4 minutes by Westbury, but a splendid climb to Brewham summit was halted in its tracks at Brewham Box and we then crawled over the six miles to Castle Cary, taking over 20 minutes. We got clear after that but running was restrained as something was not far ahead of us. The 106 miles from Reading to Taunton were run in 110 minutes net.

I did eventually get my run with a 47XX but not on that route. I'd gone down to Swindon on the Friday before Whitsun Bank Holiday with 5066 *Sir Felix Pole* on a relief to South Wales and got 4708 on a relief from Gloucester which it took over from a new 'Hymek' at Swindon. The Swindon driver, unused to passenger work and schedules, had allowed 4708 to accelerate to 80mph at Maidenhead before becoming concerned at the engine's rolling. In a conversation at Paddington when I gently reminded him of the 47XX maximum allowed speed of 60mph, he sought my advice for he was to work a relief back with a very rough Swindon engine, 5002 *Ludlow Castle,* which he knew from freight turns he'd worked on. He had no idea of the speeds required so my advice was to go hard to West Drayton to get up to around 65mph and then stick at that speed. That should be more than adequate for any relief train timing!

It is interesting now to see from my notebooks just how many Saturday afternoon or weekday evening runs I managed in the year before I became a 'Traffic Apprentice' in the Autumn of 1961. Punctuality was overall poor, but I show actual start and finish punctuality, where known, and the net gain (+) or loss (-) by crew and engine. My runs, excluding the many LMR and ER ones, were:

9.3.60	5079 1.40pm Paddington-Reading/Bristol, T – 9 L (+1)	
24.5.60	5051 3.5pm Swansea/Reading-Paddington, T - T (+4)	
1.6.60	4081 1.40pm Paddington-Reading/Bristol, T – 3¾ L (+ 5)	
	5059 12noon Cardiff/Reading-Paddington, 10 L – 13 L (+ 3)	
	7028* 2.40pm Neyland/Reading-Paddington, 6 L – 4 L (+ 3½)	
20.6.60	5078 1.40pm Paddington-Reading/Bristol, T – 1 L (+3)	
24.6.60	5087 5.43 arr ex-Bristol/Reading-Paddington, 3½ L – 3½ E (+7)	
3.9.60	6965 12.05pm SO Paddington-Newton Abbot/Kingswear, 4 L -12 L (+17)	
	D604 12noon Penzance/Newton Abbot-Exeter, 2 L – 2 L (+ ½)	
	5993 4.35pm SO from Kingswear /Taunton-Paddington, 8 L - 54½ L (+7)	
8.9.60	5088* 6.23pm FO Paddington-Banbury/Wolverhampton, T - 7 E (+11)	
	6029 4.30pm Birkenhead/Banbury-Paddington, 3 L – 8½ L (+2)	
16.9.60	6004 5.55pm *Red Dragon* Paddington-Swindon/Swansea, T - 1 E (+9)	
	5074 2.30pm Neyland/Swindon-Paddington, 4 L - 16½ L (+ 4)	
18.9.60	6961 12.05am Paddington-Wolverhampton, 4918 to Chester (overnight)	
20.9.60	5011 8.55pm (Sun) Birkenhead/Wolverhampton-Paddington via Oxford (overnight)	
21.9.60	4095 5.30am Paddington-Reading (not timed)	
7.11.60	7035 5pm *Cheltenham Spa Express* Paddington-Kemble/Cheltenham, T - 2½ L (+8)	
	6986 5.38pm Cheltenham/Kemble-Swindon, T – 2E (+2)	
	5006 2.30pm Neyland/Swindon-Paddington, 24½ L - 21½ L (+9)	
17.11.60	7018* 7.30am Paddington-Bristol, T – 7 L (+9)	
	5095* 8am Plymouth/Bristol-Shrewsbury/Liverpool, 3½ L - 13½ E (+44)	

	6922 4.30pm Birkenhead/Shrewsbury-Wolverhampton, 3 L – 5 L (- 3)
	6024 4.30pm Birkenhead/Wolverhampton-Paddington, 5 L - 12 L (+11)
28.11.60	5048 5.55pm *Red Dragon* Paddington-Swindon/Swansea, T – 26 L (+2)
	7001* 2.30pm Neyland/Swindon-Paddington, 12 L – 20 L (+5)
- .12.60	7004* 6.45pm Paddington/Reading-Oxford/Worcester, T – 2E (+ 2)
	5088* 5.50pm Hereford/Oxford-Paddington, 24 L – 29 L (+ 8½)
8.2.61	5033* 6.5pm Paddington-Reading/Oxford (not timed)
	7023* 5.50pm Hereford/Reading-Paddington (not timed)
13.2.61	6005 6.10pm Paddington-Wolverhampton/Birkenhead, T – 11 L (+10)
	6915 6.10pm Paddington/Wolverhampton-Shrewsbury, 12 L – 16 L (-)
14.2.61	73093 Manchester/Shrewsbury-Newport (Parcels), (overnight)
	4080* Manchester/Newport-Cardiff, T – T (overnight)
	6018 1.13pm Cardiff-Pontypool Rd/Crewe, T – 1 ½ E
	6018 4.13pm Crewe/Shrewsbury-Newport/Plymouth, 24½ L – 31½ L (+4)
15.2.61	4086 8.55am Cardiff-Pontypool Road/Manchester, T – T (+ 2)
	4037 8am Plymouth/Pontypool Road-Shrewsbury (footplate)/Liverpool, 10L– T (+ 29)
	4086 12.15pm Manchester/Shrewsbury-Pontypool Road/Plymouth, 87 L – 95 L (+11½)
16.2.61	5022* 8.55pm Birkenhead/Shrewsbury-Wolverhampton (not timed)
	4075 8.55pm Birkenhead/Wolverhampton-Paddington via Oxford (overnight)
11.4.61	6169 5.42pm Paddington-Bourne End, T – T
	1453 Bourne End-Marlow-Bourne End (footplate)
15.4.61	7000 1.40pm Paddington/Swindon-Stroud, 5967 to Swindon, 9 L – 5 L (+ 12)
	7037 4pm Cheltenham/Stroud-Paddington, 6 L – 4 L (+14)
17.4.61	5043* 7.30am Paddington-Bristol, T – 11 L (Pdn – S'don net – 8, Sdon – B'tol +5)
	5024 8am Plymouth/Bristol-Shrewsbury/Liverpool, 12 L – 90 L (diverted via Glos)
	(Hereford-Shrewsbury +10)
	5971 4.30pm Birkenhead/ Shrewsbury-Wolverhampton, 1½ L – 4 L (+2)
	6002 4.30pm Birkenhead/Wolverhampton-Paddington, 5 L – 26 L (+1)
22.4.61	7028* 10.55am *Pembroke Coast Express* Paddington/Cardiff/Pembroke Dock, T – 17½ L (2½ L, WTT
	retimed for SO overload) (- 5 Public book, +10 WTT)
	5014 1.5pm *Pembroke Coast Express*, Pembroke Dock/Newport-Paddington, 2 L – 15 L
	(Spl stop Badminton horse trials, net non-stop, +6)
18.5.61	6006 5.10pm Paddington-Banbury/Wolverhampton, T – 2 L (+5)
	5036* 2.35pm Birkenhead/Banbury-Paddington, T – ½ E (+12)
25.5.61	5047 5.10pm Paddington-Banbury/Wolverhampton, T – 6 L (- 3, 142 tons overload, 465 tons on 4star
	timing after failure of 6011 on up run)
	6025 2.35pm Wolverhampton/Banbury-Paddington, T – 2 E (+7)
27.5.61	5006 2.30pm Neyland/Cardiff-Paddington, T – 10 L (+10)
3.6.61	7027 12.45pm Paddington-Evesham/Hereford, T – 3 L (+12 Oxford-Evesham)
	7006* 1.50pm Hereford/Evesham-Paddington, 5 L – T (+15)
	4074* 7.5pm Paddington-Reading/Gloucester (not timed)
10.6.61	4078 9.18 SO Paddington-Paignton, 2 L – 17½ L (+ 6)
	4930 1.50pm Kingswear/Exeter-Paddington, 4165 from Paignton, 13½ L – 5 L (+15)
24.6.61	5056* 8.25am SO Paddington-Plymouth, T – 31 L (estimated +30 after severe delays)
	5075 3.5pm SO Paignton/Exeter-Taunton/Wolverhampton, 25 L – 29 L (+4)
	4077 4.35pm Kingswear/Taunton-Paddington, 9 L – 11 L, (+10)
1.7.61	7037 12noon SO Paddington-Exeter/Plymouth, 1 L – 5 L (+21)
	6855 3.5pm SO Paignton/Exeter-Taunton/Wolverhampton, T – 5 L (+2)
	D801 4.35pm SO Kingswear/Taunton-Paddington, 15 L – 5 L (+15)

15.7.61	6960	11.15am SO Paddington-Gloucester, 5 L – 16½ L (+4)
22.7.61	5017	11.45am SO Gloucester-Paddington , T – 1 E (+22)
29.7.61	5008*	12noon SO Paddington-Taunton/Plymouth, 10 L – 24 L (+15)
	6372	10.50am Wolverhampton/Taunton-Dulverton, 46 L – 43 L (+5)
	2240	2.55pm Ilfracombe/Dulverton-Taunton, 8 L – 1 E (+7)
	D843	4.35pm SO Kingswear-Paddington , 10 L – 13 L (+6)
4.8.61	5066*	1.47pm Relief Paddington-Swindon/Swansea, T – 6 L (+15)
	4708	Relief Gloucester/Swindon-Paddington, 1 L – 3¾ L (+ 7½)

Many of the above runs were on summer Saturdays when congestion on the holiday routes to the West of England made punctuality a remote possibility. The best engine performances above were with 5087 (24 June 1960), 5088 (8 September 1960), 5006 (7 November 1960), 5095 (17 November 1960), 4037 (15 February 1961), 7037 (15 April 1961), 5036 (18 May 1961), 5056 (24 June 1961), 7037 (1 July 1961) and 5008 (29 July 1961). The two runs on the 7.30am Paddington were very curious affairs when change of driver at Swindon transformed performance, 7018 from excellent to ordinary and 5043 from poor to superb! 5047 covering a failure on the 5.10pm Paddington was in poor condition but worked flat out and maintained steam losing surprisingly little time. No Castles had time to set against the engine apart from 5043 and 7028 for parts of their journey and it was evident that it was driver rather than engine performance on both occasions.

View from the 8am Plymouth-Liverpool headed by 5024 *Carew Castle* and Gloucester's 6941 *Fillongley Hall*, diverted via the Gloucester-Hereford line because of signalling installations overrun at Maindee Junction, Newport, passed by 5095 *Barbury Castle* on the 9.10am Liverpool-Plymouth, 17 April 1961. The Shrewsbury men on 5095 chose not to wait for an assisting engine over the diversionary route. (David Maidment)

Shrewsbury's regular engine on the 8am Plymouth-Liverpool double-home diagram, 5095 *Barbury Castle*, at a very wet Pontypool Road, 17 November 1960. It was waiting time and ran consistently early arriving at Shrewsbury nearly a quarter of an hour before time. (David Maidment)

Newton Abbot's 5024 *Carew Castle* draws into Bristol Temple Meads with the 8am Plymouth-Liverpool, as D828 waits with a Paddington train, 17 April 1961. (David Maidment)

Old Oak Common's 5056 *Earl of Powis* piloted by North British diesel hydraulic D6327 from Newton Abbot with the 8.25am (SO) Paddington-Penzance at Plymouth North Road, 24 June 1961. (David Maidment)

4077 Chepstow *Castle* was a nice surprise on the 4.35pm Kingswear-Paddington on 24 June 1961. It is seen here at Cholsey two days later with a Bristol-Paddington train, 26 June 1961. It has been provided with a 4-row superheat boiler and mechanical lubrication late in the day and like 4074, retained its original front end frame but with a narrow box inside cylinder casing but no double chimney. It was withdrawn in August 1962. (MLS Collection)

Landore's 5006 *Tregenna Castle* at Cardiff General awaiting departure with the Up *Pembroke Coast Express*, 6 March 1961. It hauled the 2.30pm Neyland a couple of months later on 27 May when I was in an additional coach provided for the London Division Passenger Staff Office annual outing to Weston-super-Mare and paddle-steamer to Cardiff. I was alone I think in showing any interest in the running as the beer was flowing pretty freely by then. (John Hodge)

The run with 4037 on the 8am Plymouth-Liverpool was made on the footplate as part of the preparation for an article I was intending to write on the North & West route and my log was published in O.S. Nock's book, *The GWR Stars, Castles and Kings*. I show the full log and working in the table below:

4037 *The South Wales Borderers* 83A

11 coaches, 386/410t

15.2.61

Driver D. Lewis, Fireman R. Aggett. Insp H. George (Newton Abbot) **(Newport)**

Miles	Location	Times	Speeds		Working	Gradients
0	Pontypool Road	00.00		10L	210psi	
	Little Mill	-	50/63		½ regulator	1/104 F
4.1	Nantyderry	06.33	pw 30*			1/80 F
6.7	Penpergwm	09.45	59/52*		2nd port opened	1/153 R
	Abergavenny	16.05	pw 10*	12L	Full reg 35% cut-off	1/85 R

4037 *The South Wales Borderers* 83A

11 coaches, 386/410t

15.2.61

Driver D. Lewis, Fireman R. Aggett. Insp H. George

(Newton Abbot)   (Newport)

Miles	Location	Times	Speeds		Working	Gradients
10.4	Abergavenny Junction	18.13	34/28		220 psi	1/82 R
13.4	Llanvihangel	25.00	30/36/40	11L	30% cut-off, 225 psi	1/95 R
15.8	Pandy	27.45	67/pw 48*		½ regulator 210 psi	1/100 F
20.9	Pontrilas	33.01	68/60*	11L	shut off	1/170 F
	St Devereux	36.48	64		½ reg 15%, 210 psi	
26.7	Tram Inn	39.18	58/60		shut off	
30	Red Hill Jcn	44.41	pw 15*/62	13L	¾ reg 25% cut-off	
	Rotherwas Jcn	46.45	70		205 psi	1/92 F
<u>33.4</u>	<u>Hereford</u>	<u>49.05</u>		<u>11L</u>	195 psi	
0		00.00		9L	220 psi	
	Barrs Court Jcn	03.05	sigs 5*		Full reg	
1.6	Shelwick Jcn	04.50	48/56	8 ½ L	½ reg, 15% cut-off	
4.2	Moreton-on-Lugg	07.46	60		210 psi	L
7.5	Dinmore	11.21	56/53		½ reg, 20% cut-off	1/100 R
10.2	Ford Bridge	14.33	67		15% cut-off	1/300 F
12.6	Leominster	16.45	62/64	9L	220 psi	L
15.7	Berrington & Eye	19.48	65/60		Pricker used on fire	
18.9	Woofferton	22.47	71	8L	¼ reg 200 psi	1/100 F
23.5	Ludlow	27.16	60	6 ½ L	½ reg, 215 psi	1/112 R
	Bromfield	29.30	61		No water from troughs	L, 1/160 R
28.1	Onibury	32.03	62/51		½ reg, 17% cut-off, 218 psi	1/112 R
31.1	Craven Arms	35.26	60	5L	205 psi	
	Winstanstow Halt	-	53		Pricker used	1/105 R
	Marsh Farm Junction	-	47/ dist sigs on			1/130 R
35.6	Marsh Brook	<u>42.26//44.01</u> Xing failure			218 psi, full reg, 25% cut-off	1/112 R
	Little Stretton	-	30/35		½ reg, 20% cut-off, 200 psi	1/112 R
38.2	Church Stretton	49.58	51	8L	reg just cracked, 195 psi	1/252 R
	All Stretton Halt	-	65		shut off, 190 psi	

4037 *The South Wales Borderers* 83A

11 coaches, 386/410t

15.2.61

Driver D. Lewis, Fireman R. Aggett. Insp H. George

		(Newton Abbot)		(Newport)	
Miles	Location	Times	Speeds	Working	Gradients
41.7	Leebotwood	53.40	73		1/100 F
44.6	Dorrington	56.13	64/61	reg slightly opened	1/90 F
46.7	Condover	58.18	59	shut off, 180 psi	1/279 R, 1/134 F
	Sutton Bridge Jcn	62.26	sigs slight		
	Coleham	-	sigs 5*/pw 5*		
<u>51</u>	<u>Shrewsbury</u>	<u>66.58</u> (58.5 net)		3L	(RT Public book)

Coal – dust & ovoids, water consumption, 1,400 gallons Hereford-Salop = 28 gallons per mile, coal around 30-35 lb per mile, extremely light (estimate by Inspector George).

4037 was the regular engine from the Newton Abbot end for nearly nine months as none of the other Newton Abbot Castles was said to be fit enough for this arduous turn.

When 4037 eventually 'retired' from the run at the end of the summer, Newton Abbot borrowed Laira's 4087 for a few weeks until the turn was dieselised.

4037 *The South Wales Borderers* drops back to Shrewsbury shed after my footplate trip from Pontypool Road on 15 February 1961. The Newton Abbot engine had worked through from Newton Abbot on the 8am Plymouth and was in excellent condition, being photographed on this train months earlier (see colour section) and I saw it still working this turn in April 1961 at Bristol. (David Maidment)

I commenced my three year 'Traffic Apprenticeship' course in September 1961 and spent the first year learning basic operational management in the Western Region's London Division at Maidenhead station, Slough Goods, South Lambeth Goods, Oxford and Reading stations before being allocated three months at Old Oak Common Motive Power Depot. I've already referred in a previous chapter to my commuting from my Reading lodgings to Slough and Oxford, though I reserved for special mention here one thrilling run I had in April 1962 on the 5.30pm Oxford-Paddington booked in the hour. The engine was double chimney 5001 and the full log is given below.

5.30pm Oxford-Paddington, 18.4.1962
5001 *Llandovery Castle* – Old Oak Common
6 chs, 188/200 tons

Miles	Location	Times	Speeds	
0	Oxford	00.00		T
5.1	Radley	06.48	62/72	
7.3	Culham	08.37	78/82	
	Didcot North Junction	10.40	58*	
10.6	Didcot East Junction	11.35	48*	
14.9	Cholsey	15.48	66/79	
	Goring	18.38	85	
21.9	Pangbourne	21.01	88	
	Tilehurst	23.04	90	
27.4	Reading	24.53	86	¾ E
32.4	Twyford	28.27	82/88	
39.2	Maidenhead	33.13	90	1¾ E
	Taplow	34.27	92	
44.9	Slough	37.12	90/88	2¼ E
50.2	West Drayton	40.50	87	
54.3	Southall	43.48	88	2¾ E
57.7	Ealing Broadway	46.12	90	
	Acton	47.13	82	
	Old Oak Common E.	48.32	70	
62.1	Westbourne Park	50.05	sigs 20*	
<u>63.4</u>	<u>Paddington</u>	<u>53.09</u>		<u>6¾ E</u>

I observed the same train passing Twyford at high speed the following day behind single chimney 4082 *Windsor Castle* (the original 7013) and by my watch it was a good minute earlier. Rumours were going round Old Oak that 4082 had got to Paddington in 51 minutes, an average of 76 mph.

I started my three months at Old Oak Common for my motive power training in April and had moved into the Twyford stationmaster's house with Bob Poynter (a former

Traffic Apprentice) at the end of the month for three weeks' intensive footplate work, although I'd been given an 'all stations' WR footplate pass by the shedmaster, Ray Sims, who remembered my enthusiasm and service back in 1957 and 1958. During the three weeks appointed for footplate work observation I had to experience fast passenger steam and diesel, stopping passenger steam and DMU, fast freight, diesel shunting at Acton yard, empty stock working to Paddington and heavy unfitted freight. Outside that three week period, I had some training in the depot but late afternoons and evenings I could please myself and used the pass to go to Banbury, Swindon, Oxford and Reading on many occasions. Much of my express passenger work was on 'Kings' on the Wolverhampton road and on the North & West with Canton's 6018, and I was obliged to go to Plymouth and back on the brand new D1001 (but it failed the previous day and I got a North British Warship that leaked fumes so badly that I baled out at Exeter and breathed in the ozone in the cab of 4909 *Blakesley Hall* on the Kingswear portion).

I finished my first Saturday morning session at Old Oak on 28 April and promptly went to Paddington clutching the footplate pass I'd been handed earlier in the week, wondering where to go. I saw a resplendent 7031 *Cromwell's Castle*, recently transferred from Laira to Worcester, back down onto the 1.15pm to Hereford and with some trepidation, approached the crew. The Worcester men waved me up and we had an exemplary run to Oxford in 68 minutes arriving three minutes early without any sense of pressure. The train was 9 coaches, 306/325 tons and we accelerated smartly from Paddington and with ¾ regulator and 15 per cent cut-off sustained 68-70mph from West Drayton to Maidenhead, then the driver pulled back to just the first port and we bowled along at 65-66 mph until the Didcot turnout. There was initial heavy firing before leaving Paddington – the coal was slack and rather slaty – and the fire had been run down whilst standing at Ranelagh Bridge. The fire was built up at the back end, and after a slight signal check at Southall, firing was light. On arrival at Oxford, I saw Worcester's 7013 *Bristol Castle* run in with the 12.5pm Hereford and decided to take it back, keen to experience one of the 40XX (it was of course the former 4082) with double chimney and 4-row superheat. 7013 had an identical nine coach load and gave me a splendid run. We left Oxford on time but had a long 20mph p-way slack before Didcot, though we accelerated after Didcot to 75mph by Pangbourne with ½ regulator and 17 per cent cut-off. We were a minute early into Reading but spent four minutes overtime there with GPO mail traffic which gave an incentive to work a bit harder on the last leg. We cleared Maidenhad in 13 minutes at 75 mph and with ½ regulator and 17 per cent cut-off kept at 74-75 mph all the way to Acton. A fifteen second stand at Ladbroke Grove threatened our recovery but we got away crisply and arrived at Paddington's platform 9 in 38¼ minutes (36½ net) exactly on time. 7013 was strong and rode well, though not quite as smoothly as 7031. Firing was very light and pressure remained between 215 and 225lb psi all the way. I was impressed.

Both drivers impressed on me that engines did not always ride so smoothly and easily as I'd experienced on these trips. As I was meant to be developing an understanding of staff views and attitudes as well as the railway mechanics and procedures, I thought I ought to experience a rough engine. A run-down 'Hall' on the 4.35pm Paddington-Didcot commuter train looked a likely candidate but 4917 *Crosswood Hall* was steady enough – just felt like a bicycle with a flat tyre. I saw 5008 *Raglan Castle* ready to leave the depot for an evening commuter train a couple of days later and remarked to someone that she's a strong engine (it had received a double chimney the previous summer and I'd had a good run to Taunton as far as the signalmen would allow). 'Maybe, but she's rough,' opined a passing engineman, so I decided there and then to give it a go. I was welcomed by Driver Marshall and Fireman Stares, though I didn't tell them why I'd chosen to ride with them. I peered into the firebox, the fire looked pretty black and dead and there was only 160lb on the steam gauge, but the crew didn't seem perturbed. We hooked up to ten full coaches, 312/350 tons that formed the 5.38pm Paddington-Didcot, stopping at Twyford, Reading and then all stations. While waiting to depart, I leaned against the side of the cab and gave myself a scare because it moved. A few rivets were missing and a week later 5008 was removed from a milk train after the crew complained that the cab was

Worcester's 7031 *Cromwell's Castle* romps into Paddington with the 10.5am Hereford in May 1962, a few days after I footplated it on the Saturday 1.15pm to Oxford. (A.C. Gilbert/MLS Collection)

Worcester's 7013 *Bristol Castle* (ex-4082) at Oxford on 6 May 1962, a week after I returned to Paddington on its footplate. Rebuilt front end, valveless mechanical lubricator and double chimney are very clear in this photograph. (L.W. Perkins/F.K.Davies & John Hodge Collections)

unsafe. We got the 'RA' indicator and still had only 170lb on the gauge. The driver opened her up and without a slip we marched out of the station and with full regulator and 20 per cent cut-off we shook, rattled and banged along, every bit of metal in the cab doing a war dance. We passed Southall in twelve minutes, already doing 68 and although 5008 was rough it was predictable, the engine snaked, pitched and rolled but not violently and I soon got used to the rhythm, finding it quite fun. There was no doubt the engine was strong. The blast was livening up the fire and after a good stirring with the long pricker by Stares around West Drayton the pressure began to rise and by Slough we were sailing along at 72mph, cleared in 21 minutes and the driver eased back the regulator and wound the reverser up to 17 per cent. Speed fell to 67 and we crossed to the relief line at Ruscombe and drew into Twyford platform four minutes early in 35 minutes for the 31 miles. While standing there, the engine began to blow off steam with the needle on 225lb psi. I discovered back on shed that its current mileage was 56,000. I tried in vain to find a really rough engine – even high mileage engines ran relatively smoothly although they rattled a bit and it was not until I rode a 'Royal Scot' (46166) on the 7.30am Penzance from Pontypool Road to Shrewsbury that I found a bad one. The engine kept time without any problem, no real effort was needed as the schedule was slack but the riding over 50mph was completely unpredictable and in danger of throwing me right across the cab with sudden lurches.

In contrast I went back to Twyford on the 5.38pm Paddington a couple of days later on 4088 *Dartmouth Castle* of Swindon depot. Like 5008, it had 4-row superheat and double chimney and I'd had a good run with it on one of the Herefords a few months previously. We had no problems with steam supply, but the driver took things very easily taking 14½ minutes to pass Southall at 57 mph, with just one third regulator and 16 per cent cut-off. Then we were cautioned at Iver and warned that there were cows on the line so crawled at walking pace to Langley. 'Now we'll get going', exclaimed the fireman, but the driver pushed the engine no harder than before. The safety valves were erupting furiously, and the fireman was yelling at his driver to open her up, 'you're wasting my f**** effort!' I thought the fireman was going to yank the regulator over, but the driver notched up to 18 per cent which got us through Slough at 58 and eventually 66 at Maidenhead, and we were three minutes late into Twyford in 42 minutes. The driver was quite happy that the delay could be explained, but I felt (and the fireman did too) that we could so easily have been on time.

After a day in one of the shed offices, I would take a trip in the evening out to Reading and back and three times came back with the 12-coach 440ton gross load on the 12.5pm Milford Haven, another Old Oak Castle turn. 5034 *Corfe Castle* was on time and accelerated to 61 at Twyford, 70 at Maidenhead and 72 at Slough, using full regulator and 18 per cent, then 15 per cent cut-off. Pressure dropped to 170lb psi at Slough as the fire was getting dirty but attention with the pricker restored pressure and it was back to 225 by Acton where we were up to 72mph again having lapsed to 64 at West Drayton. Despite a check after Acton, we arrived on time. 5056 *Earl of Powis* with the same load made it look easy – it was recently ex-works and ran smoothly at 68-70 mph without any fuss and arrived early. On the third trip, we had another Old Oak Castle in excellent condition, 5065 *Newport Castle*, and after a p-way slowing costing us four minutes at Twyford, I looked forward to harder work to recover time. But the driver was having none of it. We dribbled along at 60mph on first port of the regulator and 15 per cent cut-off and were duly four minutes late into London, the driver quite content, but with a fine steam-tight engine riding beautifully and steam sizzling through the safety valves most of the way, we could have regained all the lost time.

Most of the express passenger work during the extensive three weeks of full day footplate experience was on the Birmingham road with 'Kings' but I did spend a day sampling Castle work on a couple of Old Oak Swansea turns. On 10 May I picked up the 7.55am Paddington which ran into Reading on time behind 7036 *Taunton Castle*, an engine which had been used on much of the depot's top link work over the years, but was now some fifteen months since last overhaul, with over 60,000 miles run. I joined the cheerful Canton crew and I soon found out that they were happy to go quite hard even though our load was only seven coaches, 238/255 tons. The driver used full regulator throughout and adjusted the cut-off and we were doing 72mph by Pangbourne, winding back from 20 to 18 per cent with 75 at Cholsey (15 per cent) and 77 at Moreton Cutting before being brought to a stand by signals at Didcot station in 18½ minutes for the 17.1 miles. Before the fireman could make enquiries as to the cause of the delay, the signals came off and we continued in energetic style, with 72 at Wantage Road, 75 Challow, 74 Uffington and 76 at Shrivenham and although we were checked to 10 mph on the approach to Swindon, we ran in three minutes early in 43½ minutes for the 41.3 miles. 7036 was clearly a strong engine and rode well, the only sign of her mileage being the usual vibration of the cab controls when we were working hard. Full regulator and 17 per cent cut-off produced 79mph at Little Somerford and after speed had fallen to 64mph around Hullavington, the driver wound the reversing screw to 22 per cent and the engine responded by accelerating to 66mph up the last couple of miles to Badminton summit. It was easy after that and we free-wheeled down through Winterbourne at 76 before braking for Filton Junction. 74 in the tunnel and with just a slight check through Severn Tunnel Junction, we berthed at Newport High Street in under the hour, well on time. I'd been taking speed readings from the speedometer which the driver assured me was accurate and had kept an eye on the steam pressure gauge which had never moved from the 225 psi red line. I ceased timing after that for I found it was my turn to fire and no mismanagement on my part could shift that needle from the 225 mark. The crew changed at Cardiff and the relieving Landore crew were quite happy to see me maintain charge of the firing. They were obviously in the mood to get home as we continued running early at all points and arrived at the buffer stops at Swansea High Street three or four minutes before the 12.30pm booked arrival time.

Before leaving Old Oak the previous night I'd promised Driver Ward and Fireman Thomas who were on the 6.55pm Paddington-Swansea double-home turn that I'd come back with them on the 1.30pm Swansea (11.10am Milford Haven). After a quick snack in the High St. refreshment room, I joined the crew as double chimney 5056 *Earl of Powis* backed down from Landore onto the 11 coach train, 384/420 tons. 5056 was only a couple of months out of Swindon Works and in contrast to 7036, rode in almost complete silence and as steadily as a Pullman coach. The crew were clearly totally confident in their engine, and it was with some concern and dismay that Fireman Thomas saw the steam pressure drop alarmingly to 175lb psi as we forged up the 1 in 106 steepening to 1 in 91 to Skewen at a steady 42mph. 'Something's wrong' exclaimed Thomas and he disappeared round to the front of the engine during our stop at Neath. 'Smokebox door not fully shut, tightened it a couple of turns.' We had no problem with steam after that.

Swansea-Paddington, 10.5.1962

5056 *Earl of Powis* – Old Oak Common

11 chs, 384/420 tons

Miles	Location	Times	Speeds		Pressure	Working	Gradients
0	Swansea High St	00.00		T	225		
	Landore	05.05	15/10*		205		1/106 R
	Llansamlet	08.47	44		175	¾ reg, 30%	1/91 R
5.2	Skewen	12.13	42/58		185	shut off	1/88 F
<u>7.8</u>	<u>Neath</u>	<u>15.40</u>			<u>210</u>		
0		00.00			225		
2.2	Briton Ferry	03.35	50		225	1/3 reg, 18%	L
<u>5.6</u>	<u>Port Talbot</u>	<u>09.04</u>			<u>225</u>		
0		00.00			210		

		Swansea-Paddington, 10.5.1962				
		5056 *Earl of Powis* – Old Oak Common				
		11 chs, 384/420 tons				
Miles	Location	Times	Speeds	Pressure	Working	Gradients
	Steel Works	–	38/10* sheep on track			
	Margam	04.07	53	220	½ reg, 35%	L
3.6	Margam Moors Box	–	sigs 36*	225		1/139 R
6.6	Pyle	10.39	46	225	full reg, 35%	1/93 R
8	Stormy Sidings	–	44/60*	215		1/93 R, 1/132 F
12.1	Bridgend	18.35				
0		00.00		220		
3.8	Pencoed	06.08	58/48*	225	½ reg, 30%	1/157 R
7.7	Llanharan	10.00	40*/42 colliery slack	225	½ reg, 30%	1/138 R
9.1	Llantrisant	13.29	sigs 35*	210		
	Miskin	–	58	225	¼ reg, 20%	1/187 F
13.6	Peterston	18.15	60	225		L
16.6	St Fagans	21.15	64	210	shut off	
18.1	Ely	22.44	61/sigs 5*			
20.3	Cardiff General	27.40	3 E	225		

		Signal checks, Cardiff – Newport				
0	Newport High St	00.00	3 L	225		
3.4	Llanwern	–	62			
7.4	Magor	–	68			
9.8	Severn Tunnel Jcn	–	sigs 20*			
	Severn Tunnel	–	54/77			
15.3	Severn Tunnel East	–	42	225	¾ reg, 18%	1/100 R
16.8	Pilning	–	41/49			1/100 R, L
20.3	Patchway	–	45/38	225	¾ reg, 22%	1/90 R, 1/80 R
	Coalpit Heath	–	55		¾ reg, 18%	1/300 R
28.8	Chipping Sodbury	–	58	225		1/300 R
33.4	Badminton	–	60	225	shut off	1/300 R, L
	Hullavington	–	73			1/300 F
43.7	Little Somerford	–	60* pws	225		
50.5	Wootton Bassett	–	62/ sigs 5*			
56.1	Swindon	67.00	5 E			

Thereafter it was so easy that I just sat back and enjoyed myself, 5056 behaving perfectly on a beautiful May afternoon. We ran steadily at 68-70mph unchecked to Reading, waited time and arrived in Paddington 5 minutes early without breaking sweat. With 5056 in this condition, we could have cut twenty minutes off the Cardiff-Paddington schedule with ease. I rode back to Twyford with a 'Modified Hall' (7906) on a commuter train and somewhere around Burnham passed the next Up Swansea (the 12.5pm Neyland), running very early also with Laira's 4087 (!). I almost wished I'd waited to have a run with my favourite Castle, but I couldn't let the Old Oak crew down, having promised that I'd join them. I think they were keen to show off what they and their engine could do. Our passengers were happy too.

The next day I did a triangular trip from Paddington to Shrewsbury (6016 *King Edward V* to Wolverhampton and Stafford Road's 5047 *Earl of Dartmouth* on to Salop), 6018 *King Henry VI* to Bristol via the North & West and returned from Bristol to Reading on the 4.47pm Taunton, with D803 to Swindon, then transferring to the footplate of Swindon's 5068 *Beverston Castle* which attached the Gloucester portion. Now with twelve coaches, I was surprised at the ease with which we ran the 41 miles to Reading in just about even time with a maximum of 75mph through Didcot. 5068 had a double chimney and the 4-row superheat boiler and extended smokebox that had originally been with 4090 or 4093. 5047 and 5068 had been contrasts – 5047 high mileage, rugged and noisy (but plenty of steam), 5068 smooth and quiet, aping my experience with 5056, even though worked harder.

A few more days passed with 'Kings' to Wolverhampton and back (6000 and 6026) contrasting with condenser pannier 9709 on Old Oak-Paddington empty coaching stock duties, a DMU cab ride from Banbury to London, D3103 shunting Acton Yard and the highlight, Churchward 4704 on the 10.55pm Paddington Goods-Hockley/Birkenhead fast freight which I drove from High Wycombe to Banbury, returning from Snow Hill with 6015 *King Richard III* on which I was in charge of the shovel until nearly collapsing with fatigue and therefore wisely taking to

5056 *Earl of Powis* heads the Up *Capitals United Express* through Ely Mills just west of Cardiff a few days before the author's footplate trip on this engine, 27 April 1962. (Simon Lathlane Collection)

the cushions after Banbury. Two Castles at this time were stored in the paint shop, 4098 *Kidwelly Castle* and 7030 *Cranbrook Castle*, and I was curious to note activity around the latter, and on enquiring found it was being prepared for a special high-speed test run the following day. The 'Western' 2,700hp class 52s were due to replace the 'Kings' during the summer service with schedules accelerated in the autumn and the Regional Civil Engineer wanted to check what track upgrading would be necessary. He'd therefore commissioned a high-speed run to Wolverhampton and back with the 'whitewash car', the test vehicle that detected substandard track and marked it with a splash of whitewash to alert gangers on improvements required. No diesel was available for this duty and a double chimney Castle had been selected – two hours each way to Wolverhampton with two five minute stops at High Wycombe and Leamington. I was excited and asked Ray Sims if I could use my footplate pass to travel on the engine, but received a reluctant 'no', as Driver Pimm would be accompanied by two firemen and HQ Inspector Hancock – four in the cab is the maximum allowed. Someone in the office heard the refusal and whispered to me, 'Why don't you go up to Paddington and just ask the Civil Engineer if you can ride in the train?' So the following morning I made myself available at 10am and with some trepidation asked the guy who appeared to be in charge of the whitewash car if I could travel on the train, explaining that I was a Traffic Apprentice. He hesitated a moment, then said, 'Why not, just make yourself scarce in the first coach.' So I did, no-one bothered me after that and I sat in state and timed my 'run of a lifetime'. The train was a featherweight five coaches and I was reminded of the ATC tests with 5056 that the GWR had carried out in earlier years. We were to set off at 10.25 a quarter of an hour after the Birmingham 'Blue Pullman' and the two five minute stops were for pathing reasons only to ensure we had a clear road so that the track test would not be invalidated.

Paddington-Wolverhampton Test Train, 15.5.1962

7030 *Cranbrook Castle* Old Oak Common

5 coaches, 176/180 tons

Miles	Location	Times	Speeds	Schedule	Gradients
0	Paddington	00.00		T	
1.3	Westbourne Park	02.55	52	1E	
3.3	OOC West	05.53	42*	¾ E	
7.8	Greenford	10.33	75	½ E	1/264 R
10.3	Northolt Jcn East	12.32	71	½ E	
	West Ruislip	14.14	74		L
14.8	Denham	16.30	77		
	Denham Golf Course	17.17	70		1/175 R
17.4	Gerrards Cross	18.51	69		1/175 R
	Seer Green	21.15	72		1/254 R
21.7	Beaconsfield	22.38	77/82		1/225 F
<u>26.5</u>	<u>High Wycombe</u>	<u>27.39</u>		<u>1L</u>	
		00.00		T	
3.3	West Wycombe	03.43	50		1/179 R
6	Saunderton	06.48	61/67		1/164 R

	Paddington-Wolverhampton Test Train, 15.5.1962				
	7030 *Cranbrook Castle* Old Oak Common				
	5 coaches, 176/180 tons				
Miles	Location	Times	Speeds	Schedule	Gradients
8.2	Princes Risborough	09.50	62*/80	¼ E	1/88 F
	Ilmer Halt	12.06	90		1/176 F
13.6	Haddenham	13.55	96/92		L, 1/200 F
17.6	Ashendon Jcn	-	94		L
	Dorton Halt	-	90		1/200 R
20.9	Brill	18.58	94/91		1/200 F
23.9	Blackthorn	21.02	92		L
26.9	Bicester North	23.09	83	½ L	1/200 R
30.7	Ardley	26.05	77/88	½ L	1/200 R, 1/200 F
35.9	Aynho Jcn	30.18	66*	¼ L	
	King's Sutton	31.50	70		L
41	Banbury	34.45	80	¾ L	L
44.6	Cropredy	37.48	78		1/330 R
	Claydon Xing	39.50	77		
49.8	Fenny Compton	42.00	pws 66*		1/251 F
54.7	Southam Road	46.04	85		L
	Fosse Road	47.52	87		1/187 F
<u>60.8</u>	<u>Leamington Spa</u>	<u>51.45</u>	(70.5 average)	<u>2¼ L (49m 45secs net)</u>	
0		00.00		T	
2	Warwick	03.04	60/70		1/109 F
6.2	Hatton	06.56	66	T	1/95 R, 1/110 R
	Hatton N.Jcn	-	55*		L
	Lapworth Troughs	-	72		1/200 F, L
10.4	Lapworth	11.01	77		1/231 R
12.9	Knowle & Dorridge	13.14	74 eased		L
	Widney Manor	15.03	pws 16* (long)		L
	Solihull	18.56	54		L
	Olton	20.56	75		1/223 F
	Acocks Green	21.45	77		1/267 R

Paddington-Wolverhampton Test Train, 15.5.1962
7030 *Cranbrook Castle* Old Oak Common
5 coaches, 176/180 tons

Miles	Location	Times	Speeds	Schedule	Gradients
20.1	Tyseley	22.33	79/64*	4½ L	L
	Small Heath	23.28	63		
	Bordesley	24.27	59		L, 1/45 R
23.3	Birmingham Snow Hill	25.59	22* (Through Line)	4L	
24.2	Hockley	27.57	46		
	Soho & Winson Green	28.54	56		L
	Handsworth	30.06	62	1½ L	1/100 R
28	West Bromwich	32.28	60/41		1/100 R
	Swan Village	34.06	61/sigs		L
30.7	Wednesbury	37.23	sigs 2*		1/147 F
	Bilston Central	41.13	sigs 2*/54		L
34.2	Priestfield	43.07	sigs 20*/ sigs 30*/48		L
35.9	Wolverhampton LL	46.55		5L	

Five minute stops at High Wycombe and Leamington for pathing (10.10 Paddington 'Blue Pullman' diesel ahead). Signal checks from Swan Village to Wolverhampton – we had caught the 'Blue Pullman' up!

2.20pm Wolverhampton LL-Paddington
Loco, formation and crew as on down run

Miles	Location	Times	Speeds	Schedule	Gradients
0	Wolverhampton LL	00.00		T	
1.7	Priestfield	03.25	48/34*		
	Bilston Central	05.08	52		L
5.2	Wednesbury	07.52	58/43*		
	Swan Village	10.03	51		1/100 R
	West Bromwich	11.25	63		
10	Handsworth	13.45	69	1¼ E	1/100 F
	Soho & Winson Green	14.34	42*		
	Hockley	15.43			
12.6	Birmingham Snow Hill	17.08	38* (Through Line)	3E	
	Bordesley	18.58	64		1/45 F, L

Miles	Location	Times	Speeds	Schedule	Gradients
		2.20pm Wolverhampton LL-Paddington			
		Loco, formation and crew as on down run			
	Small Heath	19.58	67		
15.8	Tyseley	20.54	72	3E	
	Acocks Green	21.47	77		L
	Olton	22.34	78		
19.6	Solihull	24.02	77/82		L
	Widney Manor	25.15	74 eased		
23	Knowle & Dorridge	27.23	pws 17* (long)	2E	
25.5	Lapworth	31.54	76		1/258 F
	Lapworth Troughs	-	83		
29.7	Hatton	35.21	86/65*	1L	1/177 F, L
33.9	Warwick	38.38	84/pws 81*		1/108 F
<u>35.9</u>	<u>Leamington Spa</u>	<u>41.10</u>		<u>1¼ L (36.30 net)</u>	
0		00.00		T	
	Fosse Road	04.57	65		1/187 R
6.1	Southam Road	07.15	71		L
	Greaves Sidings	-	80		L
11.2	Fenny Compton	11.21	80/79		1/251 R
	Claydon Xing	13.28	82		
16.2	Cropredy	15.17	88		1/179 F
19.8	Banbury	17.58	80*	1L	L
	King's Sutton	20.45	84		L
24.9	Aynho Jcn	22.14	65*/72	1¼ L	
30.1	Ardley	26.32	76	1½ L	1/200 R
33.9	Bicester North	29.07	102	1½ L	1/209 F
36.9	Blackthorn	30.55	105/102		1/200 F
39.9	Brill	32.53	94		1/200 R
	Dorton Halt	34.29	87 eased		
43.2	Ashendon Jcn	35.42	61*/74	1¾ L	L
47.2	Haddenham	39.04	80		1/200 R
	Ilmer Halt	41.12	81		1/176 R
52.6	Princes Risborough	43.31	80/62*	2L	

2.20pm Wolverhampton LL-Paddington

Loco, formation and crew as on down run

Miles	Location	Times	Speeds	Schedule	Gradients
55.8	Saunderton	46.25	68		1/88 R, 1/100 R
58.5	West Wycombe	48.48	81/48*/62		1/164 F
60.8	High Wycombe	51.59	(70.2 average)	1½ L	
0		00.00		T	
4.8	Beaconsfield	05.15	76		1/225 R
	Seer Green	06.28	85		1/254 F
9.1	Gerrards Cross	08.21	94		
	Denham Golf Club	09.29	99		1/175 F
11.7	Denham	10.01	103	½ E	1/264 F
	Denham Troughs	-	101		
	West Ruislip	11.47	91	L	
16.2	Northolt Jcn East	13.11	94	¼ E	1/264 F
19.7	Greenford	14.40	97	¼ E	1/264 F
23.2	OOC West	18.25	40*/58 RL	T	
25.2	Westbourne Park	21.55	pws 15*	1½ L	
26.5	Paddington arr	25.15		¾ E	

7030 *Cranbrook Castle* races past Gerrards Cross at 70mph with the Paddington-Wolverhampton-Paddington special high speed track testing train with the author in the first coach, 15 May 1962. The track-testing 'whitewash coach' is at the rear. (Celyn Leigh-Jones/John Hodge Collection)

Although I claim 105mph at Blackthorn, I only got one reading at that speed and one before and one after at 103, so I can't be sure 105 was actually reached.

The next day was my 24th birthday and I noticed from the Old Oak chalked engine board that 7030 had gone down to Oxford on a parcels and would return with the 5.30pm fast train back, so I decided to celebrate it from the footplate. I egged on the driver telling him of 7030's exploits the previous day and he had a go, although when we were running at 90 near Maidenhead he eased, saying that he would not feel happy going faster as we had begun to roll a little.

		17.30 Oxford-Paddington			
		16.5.1962			
		7030 *Cranbrook Castle* **81A**			
		6 coaches, 188/200t			
Miles	Location	Times	Speeds	Schedule	Working
0	Oxford	00.00		T	
	Kennington Jcn	03.59	56		
5.1	Radley	06.34	70/69		
7.3	Culham	08.31	72		
	Appleford Halt	09.23	75		
	Didcot North Jcn	10.37	60*		
10.6	Didcot East Jcn	11.33	44*	T	
	Moreton Cutting	13.09	58		
14.9	Cholsey	16.00	70		½ regulator, 20% cut-off
	Goring	19.04	75		
21.9	Pangbourne	21.38	78		
	Tilehurst	23.49	82		2nd port reg. cracked open
27.4	Reading	25.46	83	¼ L	
32.4	Twyford	29.23	81/84		17-20% cut-off (varying)
36.4	MP 27	-	88		
39.2	Maidenhead	34.03	90	1 E	225 psi steady
	Taplow	35.16	88		firing very light, 17 %
	Burnham	36.18	90		
44.9	Slough	38.00	88/87	1½ E	
	Langley	39.29	85		
	Iver	40.32	88		
50.2	West Drayton	41.35	85/81		shut off momentarily
	Hayes	43.15	82		
54.3	Southall	44.34	85	2 E	
57.7	Ealing Broadway	47.13	69 sigs - double yellow		
	Acton	48.28	72		
	Old Oak Common West	49.58	sigs 16*		
62.1	Westbourne Park	53.12	47/ sigs 25*		
<u>63.4</u>	<u>Paddington</u>	<u>56.19</u> (54 mins net)		<u>3½ E</u>	

I returned to Reading on the footplate of sister locomotive 7032 *Denbigh Castle* on the 7.5pm Paddington – Gloucester and caught a DMU back to Twyford to share my experiences with my host. The following day I was aiming for a triangular trip via the North & West, so picked up the 7.55am Paddington with another Old Oak Hawksworth engine, 7021 *Haverfordwest Castle,* for many years based at Landore. We almost duplicated my experience of 7036 on the same train, but I left at Newport, caught a DMU across to Bristol to pick up the 7.30 Penzance, an ex-works 6945 *Glasfryn Hall* replacing the 'Warship' and got my 'rough' 'Royal Scot' from Pontypool Road to Shrewsbury returning home with a 'County' (1026) and 'King' (6022) on an afternoon train from Birkenhead. The day finished off with me joining the footplate of 4096 *Highclere Castle* on the 7.5pm Paddington which I fired to Reading before alighting there. This old 1926-built engine was in beautiful condition and steamed very freely so my job was straightforward and I was left to get on with it as apparently my competence at firing was rumoured round the depot. As soon as the driver closed the regulator as we passed Twyford ready for the Reading stop, the engine began to blow off steam furiously – I suppose I should have anticipated this and eased back to keep the engine quiet and not waste steam, but it would be needed after Reading as we had a good load of eleven on.

A couple of days later, my dad had a reunion with some of his old army pals in Bristol, and I booked a compartment for him and a few London-based colleagues down on the *Bristolian* and travelled in the cab of D811 *Daring*. I'd booked them back on the 4.47pm Taunton, knowing it would be steam from Swindon, so I swapped cabs there from D856 *Trojan* to Old Oak's double-chimney 5001 *Llandovery Castle,* which again gave me a beautifully smooth run up, speed in the mid-70s, free steaming and riding like a passenger coach. I'm pleased to say that we arrived in London on the dot of time and I hung out of the cab to greet my dad's friends as though I belonged there, relieved that we had not let them down.

I was now back in the shed having to learn about crew rostering, engine diagramming and industrial relations, so my further footplate trips were restricted to my morning trains up from Twyford

Old Oak Common's 7021 *Haverfordwest Castle* on which I rode from Reading to Newport at the end of May 1962, before sampling a rough 'Royal Scot' on the 7.30am Penzance from Pontypool Road to Shrewsbury. Although relatively high mileage it rode well with minimal vibration. It was withdrawn fifteen months later in September 1963. (MLS Collection)

(a couple of Stafford Road Castles, 5019 *Treago Castle* and 5063 *Earl Baldwin*, as well as a number of 'Halls') and evening excursions to Reading or Oxford. I managed to get a run up from Reading with Worcester's immaculate double chimney 7007 *Great Western* and was not disappointed. A full boiler of steam and a nice 9-coach load and the driver treated me to full regulator and after Twyford, 17 per cent cut-off. A lovely even purring sound and we were at 80 mph by Maidenhead and held 80-82 to West Drayton when the driver eased and we coasted into Paddington unchecked in just over 36 minutes, a few minutes early.

Punctuality during my time at Old Oak was much better than my 1959 and 1960 Paddington-Reading runs recounted in my earlier Castle book. There were fewer p-way restrictions and with closer adherence to the timetable, signal checks were much reduced. This showed up the slackness of many of the WR schedules for those routes still steam-hauled and at this stage, two years before the deterioration of maintenance of the Castles and other WR motive power, the Castles had a lot in hand and rarely needed pushing hard. The temptation for some drivers was to take it too easily and I had a run down on the 11.55am Paddington-South Wales as far as Reading with an old Canton favourite, 5020 *Trematon Castle*, which disappointingly dropped a couple of minutes to Reading on the easy 40 minute schedule. The driver complained that she was sluggish, but he never cracked open the second valve of the regulator and seemed content to wander along at 60-62 mph. I felt the engine just needed a few minutes of hard blast to liven the fire and get going. We were not short of steam, but the loss of time seemed to be accepted. Of course, there were four minutes of recovery time at Steventon, so I guess the train would be on time then.

My purpose in going down to Reading in the middle of the day on one occasion was to sample a heavy parcels train, the 8.55am Fishguard. That and the 8.55pm Birkenhead Parcels were notorious runners, the latter often appearing at Paddington three or four hours late. I'd had a ride on a lovely little mogul, 6309, on an Up parcels train previously but the load had been light enough to engage 4093 *Dunster Castle* running on a parallel line with the Up Fishguard boat train in quite a competitive race. The Fishguard Parcels duly arrived at Reading not too far away from booked time with Landore's 4099 *Kilgerran Castle* and a long string of bogie parcel vans weighing 444 tons, estimated at 500 tons gross. I found my host was Swindon Driver Instone whom I'd encountered the previous year on the 80mph Churchward 2-8-0. We eventually got away and were satisfied with 50mph on the relief line, needing only ¼ regulator and 15 per cent cut-off, as the driver knew we were not far behind a semi-fast DMU, and sure enough, we were checked to walking pace at Slough. The West Drayton signalman turned us out onto the main line and Driver Instone responded by opening up to full regulator and 20 per cent and with full steam pressure, 4099 shifted the load to 60 by Southall and 68 through Ealing Broadway, before easing on our approach to Old Oak Common and platform 1A at Paddington.

I completed my London Division training at the end of May 1962 and was drafted to Margam to undertake my marshalling yard experience as it was considered that Acton was too small and I was required to see a more modern yard. I therefore set off for Swansea on the 7.55am Paddington in early June, still steam worked for a couple of weeks – behind a somewhat sluggish Reading 5067 *St. Fagans Castle* the first week and a lively Old Oak Common double chimney 5060 *Earl of Berkeley* the second. After reporting to the District Operating Superintendent at 12.30pm on the Monday morning, I was informed that henceforth I would be present at 9am, so the following Sunday evening I went down on the 4.55pm Paddington with eight jam-packed coaches and Old Oak's 7033 *Hartlebury Castle* diverted via the Vale of Glamorgan because of engineering work between Cardiff and Bridgend. Thereafter I caught the night sleeper via Gloucester, arriving at Swansea around 7.30am on the Monday morning. This was Hymek haulage, though one Monday when I was not required until the afternoon, the Hymek on the 7.55am Paddington failed between Didcot and Swindon and we were rescued by a tender first Hall (6991) and then a very high mileage 6835 *Eastham Grange* from Swindon that recovered a little of the time. 5037 *Monmouth Castle* replaced a Hymek one Friday evening on my train home (11.10am Milford Haven) and managed the 13-coach train admirably until Cardiff when the steam heating pipe between engine and first coach burst and 5037 was replaced by the Cardiff General standby, 5096 *Bridgwater Castle*, which was a poor substitute for the Llanelli engine. Occasional relief trains were still steam hauled – that summer after all the regular services were Hymek hauled, I enjoyed 5075 *Wellington* on an Up relief and 5037 and 5041 *Tiverton Castle* on the 2.55pm Paddington Down, all performing competently on 8-9 coach trains.

I spent the whole of the summer and autumn in the Swansea District and spent many an evening with a colleague running out to Llanelli or Carmarthen with the West Wales portions of the 1.55 or 3.55pm Paddington, invariably behind a Landore Castle – 4081, 5013, 5030, 5078 and 7028 the most regular. Return was always on the 6.50pm Neyland-Paddington sleeper, a 12-coach heavy train booked for a Carmarthen Castle between Carmarthen and Swansea, almost entirely the exclusive territory of 5039 *Rhuddlan Castle*. It would stop at Llanelli and be banked from Gowerton to Cockett (2 miles of 1 in 50) by Llanelli's 9408, the speed usually between 25 and 30mph at the summit. In late autumn as further Castles were drafted to Carmarthen as work for them shrunk elsewhere, we found 5027, 5054 and 5098 shared the working. The West Wales work was comfortably within the scope of the Castles (frankly 'Halls' could have performed as well) and the 6.50 Neyland sleeper was the only turn where their full power was utilised.

The winter and spring of 1963 was spent training at Marland House, the South Wales Divisional Headquarters, and evening forays to Pontypool Road, Hereford or even Shrewsbury were made with another Traffic Apprentice and fellow enthusiast. The 4.40pm Cardiff-Manchester was booked for a Cardiff Castle (Canton had closed to steam so its engines moved to Cardiff East Dock) and one of 4093, 5029, 5043, 5073, 5081 and 5092 usually appeared at the head end. During the spring of 1963, three express turns each way (including the overnight trains) were rediagrammed between Bristol and Shrewsbury from 'Warship' diesel hydraulics to St Philip's Marsh Counties or Castles, as so many diesels had suffered damage during the very harsh winter, and the 2pm and 4pm Plymouth-Crewe were two of the turns which we could pick up at Pontypool Road. 4093 moved from Cardiff and 4087 from Laira to join Bristol's 5050, 5071 and 5085 and I had excellent runs with 4087 and 4093 in particular, maintaining the diesel timings though loads were comparatively light and well below the 11-12 coach trains of earlier years. Laira's sole remaining Castle, 7022, appeared on the 4pm Plymouth on one occasion, though whether through from Plymouth or just Bristol was not known.

After spending the summer acting as Assistant Terminals Superintendent at Fishguard Harbour where my main steam haulage was with pannier tank 9602 to Clarbeston Road and civilisation, I was transferred in September 1963 to the Paddington General Manager's Office for six month's training there. This was the last steam era for the Paddington-Worcester-Hereford services. Hymeks should have taken over, but diesel availability was so poor (especially of the Class 52 Westerns which had dropped below 50 per cent) that the Worcester services reverted to steam, hauled by the remaining 85A Castles, though unfortunately they were in poor condition as they had been scheduled for withdrawal in September. Worcester often borrowed one of the Oxley Castles in lieu, though the substitute was more frequently a Modified Hall – 6989, 7920 or 7928 – or a Grange – 6856 or 6877 – which had all been through Swindon Works for general overhaul much more recently.

From 1964 my Castle experiences were mainly with special trains as described in the previous chapter – the 9 May Z48 *Great Western* with 4079, 6999, 7025, 7029 and 5054, and a couple of runs on Saturday services from the West Midlands with Oxley engines. From my first run behind a Castle in 1944 to my last in BR service in 1965, I had runs behind every Castle with just the following exceptions: 100A1, 111, 4000, 4016, 4032, 4073, 4084, 5005, 5010, 5049, 5053, 7003 and 7019. In BR service I had most runs behind 4081, 5025, 5039, 7011, 7021 and 7036. I had the longest runs/mileage behind 4081, 5011, 5014, 5037, 5056, 5084, 5092, 7001, 7011, 7018, 7023, 7025, 7030, 7032 and 7036. My first Castle run was on 26 December 1944 with 4087 from Bristol to Paddington. My last Castle run on a BR timetabled journey was with 7023 on 29 November 1964 on the 12.5am Manchester between Wolverhampton and Bristol Temple Meads.

Chapter 4
THE END DRAWS NEAR – 1963-1965

The 1,700hp 'Hymek' diesel hydraulics first appeared on the scene in the latter half of 1961 and a group were allocated to Landore diesel depot from April 1962 taking over initially the heavy Cardiff-Paddington services (8am, 10am and 12noon Cardiff and return on the 1.55, 3.55 and 5.55pm Paddington) from the Canton 'Kings'. Dieselisation of the South Wales trains worked by Landore and Old Oak Common Castles followed and the last steam booked turns on the 7.55am and 2.55pm Paddington had gone by the end of June. The first 2,700hp 'Western' diesel hydraulics (class 52) were running in during the spring of 1962 and crew training between London and Chester during the summer. Most were then used on the Paddington-Birmingham/Wolverhampton services though some released 'Warships' on the West of England expresses. Some LMR Brush Type 4s (class 47) were beginning to appear on the North and West trains, though the harsh winter of 1962/3 saw heavy snow over the whole of WR territory including Devon and Cornwall and the casualty rate among the diesels was very heavy leading to a resurgence of steam working both north and south of Bristol. South would be coverage of failures, but in the spring of 1963 three northbound and three southbound North & West expresses between Bristol and Shrewsbury were diagrammed for St Philip's Marsh Castles or Counties until the summer timetable. The diminishing stud of Bristol Castles was augmented especially for these services and transfers to assist included two first rate double chimney Castles, 4087 from Laira and 4093 from Cardiff East Dock, that were seen on trains like the 2pm and 4pm Plymouth - Crewe very frequently. Laira's remaining Castle, 7022, retained as standby, was also seen on the North & West trains at this time. Castles at Cardiff East Dock also had a couple of turns to Shrewsbury, the shed using engines like 4080, 5029, 5043, 5081, 5091 and 5092.

Double chimney 4087 *Cardigan Castle* in store at Laira in the winter of 1962/3 with 7022, the only other Castle still located there. The heavy snowstorms in December and January rendered many of the diesels inoperable, and both Castles were then taken out of store, 7022 remaining at Laira as standby for diesel failures and 4087 going later in March to St Philip's Marsh to cover diagrams on the Bristol-Shrewsbury route that had reverted to steam. (David Maidment)

Cardiff East Dock's 5091 *Cleeve Abbey* on arrival at Shrewsbury with a Cardiff-Manchester train, c1963. It appears to have deputised for a diesel on the 9.10 Liverpool-Plymouth as far as Bristol the previous day given the V93 chalked on the smokebox door. (Kidderminster Railway Museum/MLS Collection)

Carmarthen's 7012 *Barry Castle* makes a rare visit to London as it races through Southall with an Up express, 18 April 1960. (A. Tyson/MLS Collection)

Canton's 4080 *Powderham Castle* reaches the summit of Llanvihangel bank with steam to spare on the 12.40pm Cardiff-Manchester, 8 September 1962. (R.O. Tuck/Rail Archive Stephenson)

4080 *Powderham Castle* at Nantyderry with the 9.30am Manchester-Swansea, 4 August 1962. The Sugar Loaf mountain can be glimpsed to the right of the bridge in the background. (R.O. Tuck/Rail Archive Stephenson)

St Philip's Marsh's rebuilt 'Star' 5085 *Evesham Abbey* on the 2.55pm Paddington-Swansea overtakes Canton Britannia 70025 *Western Star* on the Canton-Neyland milk empties, 22 May 1961. (R.O. Tuck/Rail Archive Stephenson)

Castles were more frequently seen on football specials, excursions and relief trains, parcels and freight trains, as their express passenger work was increasingly dieselised in the early 1960s. A selection of photos follow showing the variety of work that was falling to their lot. The last major overhauls of any Castles took place in 1962. After that only casual repairs were carried out.

A Down fitted freight with Oxford's 5012 *Berry Pomeroy Castle* at Wantage Road, 7 September 1961. (MLS Collection)

Carmarthen's 4081 *Warwick Castle* saunters through the middle road at Cardiff General station with an Up freight conveying steel rods and slabs, possibly from Margam to Llanwern, June 1961. (John Hodge)

Double chimneyed 4080 *Powderham Castle* pauses at Llanelli with the Whitland-Kensington milk train, 12 June 1962. (N. Harrop/MLS Collection)

Stafford Road's 7001 *Sir James Milne* enters Oxford station with an Up parcels train, 26 June 1963. It will be transferred to Oxley shed three months later when the London Midland Region take over control of the West Midlands. (MLS Collection)

Worcester's 7002 *Devizes Castle* at Neasden after arriving with a football special for Wembley, 29 April 1961. (Stephen Sumerson/Rail Archive Stephenson)

Worcester's 7007
Great Western with a Wolverhampton-Birmingham-Worcester stopping train at Swan Village, 29 June 1960. (MLS Collection)

With the severe reduction of 'King' and 'Castle' turns occurring through dieselisation in the spring and summer of 1962, September that year saw a cull of both classes, including several Castles that had only been equipped with 4-row superheater boilers and double chimneys at the previous Works overhaul. In all, twenty-five Castles were put aside in addition to thirteen that had been withdrawn in the summer as their Works overhauls became due. Only the Hawksworth 70XX were spared though this 'amnesty' did not last long and the first 70XX – 5099, 7007, 7017 and 7030, all celebrated locomotives – were withdrawn in February 1963. An increasing number were stored during the winter of 1962. The depot allocations in August 1962, just before the 'slaughter', is indicated below and engines withdrawn in September are shown with a 'w' and double chimney engines are asterisked as before (numbers before and after the September withdrawals in brackets):

Old Oak
  Common: 4082 (ex-7013), 4089, 4098, 5001* (stored), 5008*w, 5011w, 5014, 5015, 5032*w, 5034*w, 5036*w, 5041, 5056*, 5057*, 5065, 5066*w, 5093, 7006*, 7010*, 7015*, 7017, 7018*, 7020*, 7021*, 7029*, 7030*, 7032*, 7033*, 7037 (29, 28 active, 23)

Reading: 4096w, 5018, 5067w, 5076, 5085 (5, 3)

Didcot: 4074* (1, 1)

Oxford: 5025, 5033*w, 5038, 5068*w, 7008* (5, 3)

Bristol
St Philip's
  Marsh: 5040 (stored), 5049*, 5050, 5052w, 5071*, 5075w, 5094*w (7, 3 active)

Swindon: 4079, 4088*(stored), 5002 (stored), 5023 (stored) (4, 1 active)

Newton Abbot: 5042 (stored), 5055 (stored) (2, none active)

Exeter: 4037w (1, -)

Laira: 4087*, 4095 (stored), 5029 (stored), 7022* (4, 2 active),

Stafford Road: 5019*w, 5022*, 5026*, 5031*, 5045w, 5046w, 5047w, 5063, 5072, 5088*w, 5089, 7001*, 7012, 7014*, 7019*, 7024*, 7026, 7036* (18, 13)

Shrewsbury: 4090*, 5070 (stored), 7025 (3, 2 active)

Worcester: 5099, 7002*, 7004*, 7005, 7007*,7009, 7011, 7013* (ex-4082), 7023*, 7025*, 7027, 7031 (12, 11)

Gloucester: 5000, 5007w, 5017w, 5058, 5064*w, 7000, 7003*, 7034*, 7035* (9, 6)

Cardiff Canton: 4080*, 5021w, 5043*, 5061*w, 5073*, 5081, 5091, 5092*, 5096, 5097, 7016 (11, 9)

Cardiff East Dock: Canton closed September 1962, all allocation to Cardiff East Dock

Landore: (closed)

Llanelli: 4076, 4099w, 5016*w, 5020, 5037, 5074, 5080 (stored), 5087, 7028* (9, 7, 6 active)

Neath: 4093*, 5051, 5078* (3, 3)

Carmarthen: 4081, 5027*, 5030w, 5039, 5054, 5098* (6, 5)

Total : August 1962 (128)
Total : September 1962 (100 of which 89 active)

Several of the Castles at Old Oak, Swindon, Bristol (SPM), Shrewsbury, Llanelli, Newton Abbot and Laira were in store. Eleven 70XX Castles were concentrated at Worcester for the last steam worked expresses on the Western Region, the Paddington-Worcester-Hereford trains (although someone might have overlooked the interloper, 7013, which is in fact the 1924-built 4082). The Worcester Castles were kept in excellent condition both internally and externally from 1960

By the summer of 1963 with only a few more weeks of scheduled steam haulage of the Worcester line trains, 7005 *Sir Edward Elgar* in less than sparkling condition heads the Up *Cathedrals Express* past Hinksey near Oxford, 26 June 1963. (MLS Collection)

through to the beginning of 1963, but then they deteriorated as their mileage grew and by the end of the summer service 1963, when Hymeks took over the Worcester trains, most were in poor shape. Several remained to cover for the frequent diesel failures and shortages in the autumn of 1963 and the winter of 1964, but in fact the substitutes were often Worcester's 'Modified Halls' or even 'Granges' which were in better condition having been through Swindon Works more recently.

The deteriorating external condition of the Worcester Castles is seen in this photograph too, as 7025 *Sudeley Castle* backs the stock of a Worcester arrival into Royal Oak carriage sidings before being turned at Ranelagh Bridge for its back working. (MLS Collection)

The last booked steam working from Worcester shed – 7023 *Penrice Castle* with crew and shed staff photographed for the local newspaper – ready for the 11.10am to Paddington, 7 September 1963. (John Scott-Morgan Collection)

At the end of the summer service 1963, the rank of Castles was much depleted, the number remaining and their allocations were:

Old Oak
  Common: 4082 (ex-7013, stored), 4089 (stored), 4098, 5014, 5015, 5041, 5057*, 5070, 5076, 5098*, 7006*, 7008*, 7010*, 7013* (ex-4082), 7029*, 7032*, 7035* (17, 15 active)
Southall: 7020* (1)
Reading: 5002, 5018, 7011 (3)
Oxford: 5025 (Stored)
Bristol
  St Philip's
  Marsh: 4087*, 4093*, 5040 (stored), 5071*, 5085 (5, 4 active)
Swindon: 4079, 4088* Stored) (2, 1 active)
Laira: 7022* (1)
Stafford Road: 5026*, 5031*, 5063, 5089, 7001*, 7012, 7014*, 7019*, 7024*, 7026 (10) (To Oxley, 9/63)
Worcester: 7000, 7002*, 7004*, 7005, 7023*, 7025* (6)
Gloucester: 5058 (stored), 7000, 7003*, 7034* (4, 3 active)
Hereford: 5000 (1)
Cardiff
  East Dock: 4080*, 5029, 5043*, 5056*, 5073*, 5081, 5091, 5096 (8)
Llanelli: 5037 (stored), 5042 (stored), 5054 (stored), 5074 (stored), 7028* (stored) (none active)
Fishguard
  (Goodwick): 5039 (stored), 5055 (stored) (none active)
Total : (67 of which 54 active)

Two of the lone engines – Laira's 7022 and Oxford's 5025 – moved in October to join 5000 at Hereford where they had standby duties on the North & West and to London via Worcester and the occasional turn just to Worcester. 5025, which had been stored for some time, was withdrawn almost immediately but 5000 and 7022 were joined by 5054 and 5055 ex-store in November. The class 52 'Westerns' had taken over the London-South Wales services in the winter timetable 1963, theoretically releasing Hymeks for the Worcester services, but their reliability was poor and availability that autumn and winter rarely rose above 50 per cent, making Hymek availability for the Worcester services very erratic and the authorities took the decision to return the Paddington-Worcester-Hereford services to steam traction to allow Hymeks to augment the 'Westerns' on the South Wales route.

The Hereford Castles had three diagrams, one Worcester-Paddington diagram in place of the Hymeks, a North & West turn and the Hereford standby pilot engine to cover failures on either route. The North & West turn included the 9.30am Manchester-Cardiff until March 1964 when it reverted to Brush Type 4 diesel haulage. Although the regular haulage of the main North & West expresses had returned to diesel after the spring 1963 renaissance of steam, St Philip's Marsh and Cardiff East Dock Castles continued to make occasional forays in place of diesels throughout 1963 and well into 1964. The Hereford Castles were noted several times rescuing failed diesels, 5000 on 9 January 1964 piloting D853 on the 9.5am Liverpool-Plymouth and 5055 piloting another 'Warship' on the same train passing Filton Junction an hour late on 24 April 1964. As late as 5 June 5000 was piloting D838 on the 8am Plymouth-Liverpool through to Shrewsbury. 5042 *Winchester Castle* and 5056 *Earl of Powis* joined the quartet in April 1964 in place of 5054 which had been transferred to Worcester.

7022 appeared to be the best of the Hereford Castles and was the one used most frequently on the North & West and Paddington services. Rather than go through to London, a Hereford Castle and a Worcester engine (often a 'Grange') would operate three round trips a day between Hereford and Worcester with the Worcester Castles or 'Modified Halls' covering Worcester-Paddington. The Hymeks returned to the Paddington-Worcester services in May 1964, retaining steam between Worcester and Hereford until the summer timetable commenced in June when the entire service was dieselised. Hereford's Castles were transferred in June to Worcester (7022), Gloucester (5042 & 5055) and Oxley (5000 & 5056). Three of St Philip's Marsh's Castles were withdrawn in October 1963 leaving just 4093 and 5085 to act as standby engines to cover diesel failures. The depot closed in June 1964.

Hereford's 5000 *Launceston Castle,* on standby duties on 5 June 1964, has to assist ailing D838 *Rapid* on the 8am Plymouth-Liverpool, seen here at Craven Arms. (Derek Cross)

Hereford's station pilot and standby engine, 5055 *Earl of Eldon,* draws the three coaches forming the Hereford portion of the 4.5pm to Paddington which had arrived off the 11.15am Paddington into the main Up platform, 16 May 1964. The train forward will be worked by a 'Hymek' or Worcester 'Hall' or 'Grange' if the diesel is not available. 5055, despite its neglected appearance, moved to Gloucester in July but was withdrawn in September. (John Hodge)

One of Hereford's Castles, 5042 *Winchester Castle,* acting as station pilot at Hereford station, June 1964. It is clearly in dreadful external condition, but it will move to Gloucester at the end of the summer and remain there as one of the last four of the class until June 1965.
(David Maidment)

Stafford Road's 7014 *Caerhays Castle* departs from Gobowen with the 11.40am Birkenhead which it will work to Wolverhampton, 23 March 1963. It will move to Oxley when the LMR takes over the West Midlands area six months later and 'Black 5s' will infiltrate such services.
(A.C. Gilbert/MLS Collection)

Llanelli's 5054 *Earl of Ducie* outside Carmarthen shed ready to draw stock into the station, 26 June 1963. It was transferred to Worcester in April 1964 and was found to be the Castle in best condition and most likely to be capable of 100 mph on the GW *City of Truro* anniversary special in May 1964. (N. Harrop/MLS Collection)

7013 *Bristol* Castle (ex-4082) at Worcester shed reinstated after storage there in August 1963 and its transfer to Old Oak Common at the end of September. The Davies & Metcalfe valveless mechanical lubricator is very clear in this photo. The nameplates had been removed for safekeeping whilst the engine was stored. (N. Harrop/MLS Collection)

The End Draws Near – 1963–1965 • 89

During the early Spring of 1964, Castles plus loco inspectors suddenly started appearing on the 9.15am Paddington-Worcester. The new WR General Manager had agreed to a bow-out in style of the Castles, commemorating the 60th anniversary of *City of Truro's* epic 100mph, while some were still in reasonable condition. Gerry Fiennes was not one to do things by halves and he encouraged an attempt at 100mph somewhere on the Paddington-Plymouth-Bristol-Paddington triangle proposed for the high speed tour. To test the selection of Castles for the special to be run on 9 May 1964, eight Castles with the lowest mileage (under 40,000 since Works overhaul) were rostered to the 9.15am Paddington-Worcester and the inspectors instructed the driver to attempt 100mph down Honeybourne Bank. I travelled on the train in October 1963 when Old Oak Common's 7032 *Denbigh Castle* touched 100, but I think this was too early and the inspector was just entertaining a footplate guest.

In the event, the three best Castles selected were appropriately three different versions of the class, the original and unaltered 4079 *Pendennis Castle* of Swindon depot for Paddington-Plymouth, the double chimney Old Oak Hawksworth 7029 *Clun Castle* and Worcester's standard Castle of the 5013 series, 5054 *Earl of Ducie* with Hawksworth

7032 *Denbigh* Castle at Paddington at the head of the 9.15am to Worcester and Hereford on the day it achieved 100mph in the descent of Honeybourne bank, October 1963. It was selected as one of the standby engines for the 9 May special high speed runs. (David Maidment)

tender. I had been appointed at the end of my operating management training to the post of Stationmaster, Aberbeeg in the Western Valley, the previous month and should have been 'on call' that weekend, but I persuaded my predecessor who had stayed for a couple of weeks to bed me in, to cover for me so I could ride on the special. What a day out we had! Most of us were astonished to see the old 4079 back on at Paddington expecting all the engines would be the later 70XX series. 4079 was going superbly, gaining time already on the fast schedule (Old Oak Driver Alf Perfect was said to be aiming to do Paddington-Plymouth in a record 3½ hours) and was accelerating rapidly down the four mile 1 in 222 from Patney to Lavington, speed rising to 94, 96, some even made it 98, and with 100mph looking on, suddenly a shower of sparks flew past the window and all surged with an emergency brake application. We limped into Westbury where it was found that the fire made up of the highest quality Ogilvie coal had burned so fiercely that some of the firebars had melted, dropping burning coals onto the track. 4079 had to come off and Driver Perfect was in tears. They found a grubby 'Modified Hall', 6999 *Capel Dewi Hall* (which used to be one of Canton's best), and the driver promptly produced one of the highlights of the day working the unprepared 6999 up to the high 80s before stopping at Taunton to replace it with one of the standby Castles that had been strategically placed. Some in hindsight had wished that 6999 could have gone through to Plymouth, its replacement, 7025 *Sudeley Castle*, was competent enough but unfortunately not in the same condition as 4079. The other standby by engines were 7023 *Penrice Castle* at Old Oak Common, 7008 *Swansea Castle* at Laira, 7032 *Denbigh Castle* at Bristol and 7022 *Hereford Castle* at Swindon.

Paddington-Plymouth *The Great Western Special Z48*

4079 *Pendennis Castle* – Swindon

7 Chs, 245/265 tons

9.5.1964

Driver Alf Perfect, Firemen Doug Godden & Brian Green (Old Oak Common), Insp. Bill Andress

Miles	Location	Times	Speeds	Gradients	
0	Paddington	00.00		T	
1.3	Westbourne Park	02.57	48	1½ E	
5.7	Ealing Broadway	07.58	68		
9.1	Southall	11.00	70	1½ E	
18.5	Slough	18.38	80	2 E	
24.2	Maidenhead	23.25	pws 62*	1½ E	
31	Twyford	29.08	78	1 E	
36	Reading	33.18	44*	1¾ E	
	Southcote Junction	35.51	61/74		
46.7	Midgham	43.41	68*		1/440 R
53.1	Newbury	49.00	78/75*	4 E	
61.5	Hungerford	56.02	76/66*		1/264 R
66.4	Bedwyn	60.21	74/62*	5½ E	1/477 R, L
70.1	Savernake	63.48	62	6¼ E	1/145 R, 1/106 R
75.3	Pewsey	67.55	86		1/260 F
	Woodborough	70.25	90		1/255 F
81.2	Patney	72.57	91/94		1/222 F

Paddington-Plymouth *The Great Western Special Z48*

4079 *Pendennis Castle* **– Swindon**

7 Chs, 245/265 tons

9.5.1964

Driver Alf Perfect, Firemen Doug Godden & Brian Green (Old Oak Common), Insp. Bill Andress

Miles	Location	Times	Speeds	Gradients	
86.9	Lavington	75.46	96/98	7¾ E	1/222 F
91.5	Edington Box	77.30/79.25	emergency stop		
	Heywood Road Jcn	90.06	64/sigs sl	½ L	
95.6	Westbury	98.30			

The falling burning coals had also unfortunately caused the rear driving axleboxes to run hot. 6999 then covered the 47.3 miles to Taunton in 43 minutes 7 seconds with 58½mph minimum at Brewham summit, 82 at Charlton Mackrell, 80 at Somerton and around 88 at Curry Rivell Junction.

The continuation with Driver Perfect and 7025 from Taunton:

7025 *Sudeley Castle* **- Worcester**

Miles	Location	Times	Speeds		Gradient
0	Taunton	00.00		27¾ L	
2	Norton Fitzwarren	03.45	63		
7.1	Wellington	08.59	62½		1/174 R
10.9	Whiteball Box	-	48		1/80 R
15.9	Tiverton Junction	17.39	81/78		1/155 F
27.3	Stoke Canon	26.51	84/76		
30.8	Exeter St David's	30.30	pw 43*/58	30¼ L	
35.2	Exminster	34.52	77		L
38.9	Starcross	38.29	58*		L
40.9	Dawlish Warren	40.50	49*		
42.6	Dawlish	42.42	44*/54		
45.4	Teignmouth	46.28	46*/64		
50.6	Newton Abbot	52.35	25*	30 L	
	Aller Junction	54.25	48*/54		
54.5	Dainton Box	58.37	28		1/41 R, 1/36 R
59.3	Totnes	65.10	50*	28 L	
	Tigley Box	69.35	27		1/57 R
	Rattery Box	72.55	33/37	26¾ L	1/90 R
66.1	Brent	75.53	55	26¼ L	
68.3	Wrangaton	78.43	pws 18*/53		
75.7	Hemerdon	88.56	58/72	28¼ L	1/42 F
83.3	Plymouth North Road	97.43 (94 net)		28½ L	

The Great Western Special

4.20pm Plymouth-Paddington via Bristol

7029 *Clun Castle* – Old Oak Common

7 chs, 245/265 tons

Driver H.E. Roach and Firemen Bill Rundle & Bill Watts (Laira), Inspector Bill Andress

Miles	Location	Times	Speeds	Schedule	Gradient
0	Plymouth North Road	00.00		T	
	Laira Junction	04.12	52		
4	Plympton	06.43	58*/41		
6.7	Hemerdon Box	11.30	28/26	3 E	1/42 R
8.4	Cornwood	13.43	52/54		
10.8	Ivybridge	16.35	56		1/150 R
14.1	Wrangaton	21.57	pws 17*		
16.3	Brent	24.37	60	1½ E	1/243 F
23.1	Totnes	32.17	64/53*	1½ E	1/52 F
27.9	Dainton Box	39.10	48/33		1/38 R, 1/43 R
	Aller Junction	42.46	48*		
31.8	Newton Abbot	44.11	sigs 15*	3¾ E	
37	Teignmouth	52.59	55/30*		
39.8	Dawlish	57.03	50*		
	Dawlish Warren	59.06	53 ½		
43.5	Starcross	61.28	51*/58		
47.3	Exminster	65.30	63/67		
52	Exeter St David's	70.43	35*	¼ E	
55.5	Stoke Canon	74.53	68/73		1/330 R
60.4	Hele & Bradninch	79.13	77/66*		1/243 R
64.6	Cullompton	82.53	71		1/219 R
66.9	Tiverton Junction	84.52	75		1/155 R
	Sampford Peverell	86.21	78		1/207 F, 1/216 R
71.2	Burlescombe	88.27	74		1/115 R
71.9	Whiteball Tunnel	89.11	72/88		1/115 R, 1/127 F
75.7	Wellington	91.46	96/84*		1/80 F, 1/90 F
	Victory Sidings	-	90/96		1/203 F
80.8	Norton Fitzwarren	95.12	98		1/369 F
82.8	Taunton	96.31	85*/88	4 E	

The Great Western Special

4.20pm Plymouth-Paddington via Bristol

7029 *Clun Castle* – Old Oak Common

7 chs, 245/265 tons

Driver H.E. Roach and Firemen Bill Rundle & Bill Watts (Laira), Inspector Bill Andress

Miles	Location	Times	Speeds	Schedule	Gradient
	Cogload Junction	100.26	46*		
88.6	Durston	101.41	66		1/330 F
94.3	Bridgwater	106.26	81	5½ E	L
	Dunball	108.18	85		L
100.6	Highbridge	110.56	87		L
	Bleason	115.34	90		L
	Uphill Junction	116.02	92	7½ E	
	Worle Junction	117.52	92	8¼ E	L
	Puxton	118.41	93		L
115.6	Yatton	121.02	94		L
	Nailsea	123.40	90		1/334 R
121.6	Flax Bourton	125.13	77		1/146 R, 1/200 R
	Long Ashton	127.00	75		1/234 F
	Parson Street	128.36			
126.9	Bristol Temple Meads	133.00		10 E	

5054 *Earl of Ducie* – Worcester

7 chs, 245/265 tons

Driver Rigby and Firemen R. Gitsham & C. Richards (Bath Road), Inspector Jack Hancock

Miles	Location	Times	Speeds	Schedule	Gradient
0	Bristol Temple Meads	00.00		T	
	Dr Day's Bridge Jcn	02.40			
	Lawrence Hill	03.23	42		
1.6	Stapleton Road	04.18	44		1/220 R
2.5	Ashley Hill	05.33	40		1/75 R
3.7	Horfield	07.18	39		1/75 R
4.8	Filton Junction	08.46	52	¾ E	1/300 F
	Winterbourne	12.13	61		
9.1	Coalpit Heath	13.37	70		1/300 R

5054 *Earl of Ducie* – Worcester

7 chs, 245/265 tons

Driver Rigby and Firemen R. Gitsham & C. Richards (Bath Road), Inspector Jack Hancock

Miles	Location	Times	Speeds	Schedule	Gradient
13	Chipping Sodbury	17.08	73		1/300 R
17.6	Badminton	21.09	68/87	1½ E	1/300 E, L
23.4	Hullavington	25.30	94		1/300 F
27.9	Little Somerford	28.25	96		1/300 F
30.6	Brinkworth	30.13	92		1/300 R
34.7	Wootton Bassett	33.05	70*	2 E	1/300 R
40.3	Swindon	37.53	83	2 E	
46	Shrivenham	42.00	86		
51	Uffington	45.24	90/92		1/660 F
53.7	Challow	47.13	90		1/754 F
57.2	Wantage Road	49.30	90		
61.1	Steventon	52.11	92/90	2½ E	L
64.5	Didcot	54.26	89	2½ E	1/754 F
69.2	Cholsey	57.45	88		L
72.9	Goring	60.27	83		Troughs
76.1	Pangbourne	62.53	88		L
79	Tilehurst	64.59	87		L
81.6	Reading	66.55	82*	2 E	
86.6	Twyford	70.40	78*/83	2¼ E	
93.4	Maidenhead	75.36	89	2½ E	1/1320 F
99.1	Slough	79.46	85	2¾ E	
	Iver	82.22	88		L
104.4	West Drayton	83.27	87		
108.5	Southall	86.24	92	3 E	1/1320 F
111.9	Ealing Broadway	88.43	89		
	Acton	89.43	91		
114.3	Old Oak Common W	-	90		
116.3	Westbourne Park	92.13		3¼ E	
<u>117.6</u>	<u>Paddington</u>	<u>95.20</u>		<u>4¾ E</u>	

5054 was said to have been worked at full regulator and 40 per cent cut-off in the attempt to achieve 100mph below Hullavington but there was a strong side wind. Bill Thorley was the Western Region's Traction Assistant in 1964 and had been in charge of the planning and operation of the event, including the testing and selection of the locomotives involved. In an article in the April 1973 *Railway World* magazine, he said that he had just two things he would in hindsight have changed. He would have sought the advice of Sam Ell, the Research and Development Engineer at Swindon, who would have suggested mixing Markham coal with the Ogilvie Colliery coal that was used as the latter he'd found by experience burned too hot when a locomotive under test conditions was working hard for a long period. Secondly, he would not have taken *Clun Castle* off at Bristol as it was going so well. 5054 was a good engine but not warmed through and as proven as 7029 had just demonstrated. He also found out later that 4079's firebars had been porous and failed under the extreme heat from the fiercely burning coal.

4079 *Pendennis Castle* had been identified as one of the best remaining Castles in the 100mph trials down Honeybourne bank in March and April 1964. It was accordingly smartened up and allocated to a number of railway enthusiast specials in addition to the 9 May *Great Western* tour. It is at Swindon, 25 April 1964. (N. Fields/MLS Collection)

4079 *Pendennis* Castle at Hereford at the head of a Stephenson Locomotive Society special train, 25 April 1964. (MLS Collection)

4079 *Pendennis* Castle accelerates the Ian Allan *Great Western* high speed Paddington-Plymouth away from Reading West towards Southcote Junction, 9 May 1964. (Brian Stephenson)

Above left: **After the** failure of 4079 at Westbury and the hurried use of 6999 to take the high-speed special forward, it was replaced at Taunton by the standby Castle, 7025 *Sudeley Castle,* seen here on arrival at Plymouth North Road, 9 May 1964. (David Maidment)

Above right: 7029 *Clun* Castle at Laira depot ready to undertake the Plymouth-Bristol lap of the Z48 *Great Western* Ian Allan special high-speed tour, 9 May 1964. (A.C. Gilbert/MLS Collection)

Below: 5054 *Earl of Ducie* a week after its *Great Western* high-speed run from Bristol to Paddington leaving Hereford with an Oxford University Railway Society special for Paddington via Maindee East Junction and Swindon. It had touched 93mph earlier at Honeybourne and would achieve 94 at Hullavington later, 16 May 1964. (John Hodge)

5054 *Earl of Ducie* at Taunton with empty stock in the summer of 1964, a few weeks after its participation in the Bristol-Paddington leg of the 9 May high speed tour. (MLS Collection)

By the end of May 1964, the class was reduced to just thirty-seven active locomotives, and in June the allocation was:

Old Oak Common:	4080*, 5076, 7008*, 7013* (ex-4082), 7029*, 7032*, 7035* (7)
Southall:	7020*
Reading:	5002, 5098*, 7004*(3)
St Philip's Marsh:	4093*
Worcester:	5054, 5096, 7005, 7022*, 7025 (5)
Gloucester:	5042, 5055, 7003*, 7034*(4)
Cardiff East Dock:	5039
Tyseley:	4082 (ex-7013), 5014, 5091, 7014*, 7026(5)
Oxley:	5000, 5026*, 5056*, 5063, 5089, 7011, 7012, 7019*, 7023*, 7024*(10)

In August 4080, 5076 and 7005 joined 7020 at Southall for freight and standby duties. None of the WR Castles had any regular passenger duties left (apart from the 4.15pm Paddington-Banbury) and most were covering diesel shortages or failures, especially on the Paddington-Worcester service.

Regional boundaries had changed in September 1963 with the West Midlands line north of Banbury going to the LMR with the result that Stafford Road closed and its remaining Castles went to Oxley though some were later transferred to Tyseley. The LMR imported a number of Stanier 'Black 5s' from Chester to work Chester-Wolverhampton trains sharing that duty with the Oxley Castles and 'Halls'.

However, the main passenger activity left was with the fifteen Castles now under London Midland Region operation at Tyseley and Oxley, the main ex-GW depots in the West Midlands since the closure of Stafford Road. The LMR's diesel resources were stretched to cover the main weekday services and the remaining Castles were therefore left to cover the summer Friday night/Saturday morning services from the West Midlands to the Devon and Cornwall holiday resorts which they worked as far as Bristol Temple Meads. There were five booked southbound trains in the summer peak for which Oxley Castles were diagrammed:

9.50pm (FO) Wolverhampton-Newquay

6.35am (SO) Wolverhampton-Paignton

6.55am (SO) Wolverhampton-Penzance

8.00am (SO) Wolverhampton-Ilfracombe

10.5am (SO) Wolverhampton-Kingswear.

In addition there were in July and August some relief trains to Paignton, Penzance and Bristol. The Castles returned from Bristol on:

10.5am (SO) Kingswear-Wolverhampton

11.5am (SO) Ilfracombe-Wolverhampton

11.20am (SO) Newquay-Wolverhampton

12.10pm (SO) Penzance-Wolverhampton

2.30pm (SO) Paignton-Wolverhampton

I travelled on the 8am Wolverhampton-Ilfracombe on 25 July as far as Cheltenham intending to pick up the relief *Cornishman* from there to Bristol expecting another Oxley Castle, but got instead a Crewe double chimney 'Black 5' (44765). I give the outline log of the Ilfracombe train below as an example of the Castles' last express working.

Wolverhampton-Cheltenham, 25.7.1964

8am Wolverhampton – Ilfracombe

5063 *Earl Baldwin* Oxley

12 chs, 408/440 tons

Miles	Location	Times	Speeds	Gradients
0	Wolverhampton	00.00		
1.7	Priestfield	04.12	30*/35	
2.6	Bilston	06.02	47	L
5.1	Wednesbury	10.17		
0		00.00		
1.5	Swan Village	04.15	27/21½	1/100 R, 1/95 R
2.6	West Bromwich	07.48		
0		00.00		
1.5	The Hawthorns	03.21	38	L
2.4	Handsworth	04.36	45/ sigs 20*	1/100 F
	Soho & Winson Gn	06.31	sigs 15*	1/100 F
	Hockley	08.08	pws 10*	
		09.46/12.13 sig stand		
5	Birmingham Snow Hill	15.57		
0		00.00		
	Moor Street	01.55/02.52 sig stand		

Wolverhampton-Cheltenham, 25.7.1964
8am Wolverhampton – Ilfracombe
5063 *Earl Baldwin* **Oxley**
12 chs, 408/440 tons

Miles	Location	Times	Speeds	Gradients
	Small Heath	06.34	38*	
3.3	Tyseley	08.47	15*	
	Spring Road	10.50	32½	1/200 R
	Yardley Wood	13.39	45/40	L, 1/183 R
7.4	Shirley	15.47	sigs 5*	
10.2	Earlswood Lakes	21.58	41	1/230 R
14	Danzey	26.15	67/64*	1/150 F
17	Henley-in-Arden	28.53	69	1/150 F
	Wootton Wawen	30.27	75	1/174 F
	Bearley North Jcn	32.49	48*	
22.4	Wilmcote	34.15	46/54	1/180 R
<u>25</u>	<u>Stratford-on-Avon</u>	<u>37.55</u>		
0		00.00		
3	Milcote	05.24	51	
8	Honeybourne East Jcn	11.44	52/48*	1/143 R
10.3	Weston-sub-Edge	14.20	58/55	1/202 F, 1/245 R
13	Broadway	17.31	53/pws 15*	1/200 R
17.5	Toddington	26.26	47	1/150 R
23.5	Gotherington	-	60	L
25	Bishop's Cleeve	-	67	1/150 F
26.8	Cheltenham Racecourse	37.00	62	1/200 F
<u>29.1</u>	<u>Cheltenham (Malv. Rd)</u>	<u>39.19</u>		

Apparently 5063 continued well until the outskirts of Bristol when it joined a southbound queue, though not as bad as would be found on a summer Saturday on the M5 motorway today.

At the end of 1964, the number of Castles had been reduced to just twelve. The last 1923-27 engine with original frames, 4079, was withdrawn after being taken off the 9 May special at Westbury. The last two engines of that batch, both built in 1924, but since rebuilt with new front ends, 4-row superheat boilers and double chimneys, went in August, 4080, and February 1965, 7013, the former 4082, by this time in a very rough condition. 4080 had the highest mileage of any Castle other than a rebuilt 'Star', 1,974,461, averaging 48,800 miles a year. On 1 January 1965, Tyseley had three, 5014, 7013 and 7014, and Oxley five – 5063, 7011, 7019, 7023 and 7024. All were withdrawn in February, leaving four, all based at Gloucester, 5042, 7022, 7029 and 7034, and all except

7029 in external poor condition. 7029 was in much demand for railtours and hauled the last steam train from Paddington in June 1965, the 4.15pm to Banbury, and later in November a 'Western Region Farewell' to steam from Paddington to Bristol Temple Meads and back to Swindon. The other three Gloucester based engines were withdrawn in June 1965.

Oxley's 5063 *Earl Baldwin* at Standish Junction crossing from the ex-Midland Bristol-Gloucester line to the GW line to Cheltenham and Stratford-on-Avon with the 11.5am Ilfracombe-Wolverhampton Low Level, 1 August 1964. (R.O. Tuck/Rail Archive Stephenson)

Oxley's 5056 *Earl of Powis* at Standish Junction with the 10.05am Wolverhampton Low Level-Paignton, 1 August 1964. (R.O. Tuck/ Rail Archive Stephenson)

Oxley's 7019 *Fowey Castle* at Swindon Works after a 'casual repair', 2 May 1964. The painting of the smokebox but not the rest of the engine was a clear indication of a locomotive that had received a 'casual repair' at Swindon. (G. Shuttleworth/ MLS Collection)

Old Oak Common's 5076 *Gladiator* at Oxford shed with a 'Hall', pannier tank, a Crewe 'Britannia' and a filthy Stafford Road Castle in the background, 9 June 1963. (MLS Collection)

Double chimneyed
7020 Gloucester Castle was the first Castle to be allocated to Southall as early as September 1962. It lasted two years there, being withdrawn in September 1964. A small number of other Castles were transferred there just before withdrawal, but none stayed as long as 7020. It is seen here at Didcot, 3 May 1964. (MLS Collection)

Stafford Road's
7024 Powis Castle at Oxford on 18 August 1963 just three weeks before transfer to the LMR at Oxley depot. Regrettably all the Stafford Road Castles suffered external neglect in the last months as staff numbers ran down with the impending closure of the depot, but they fared no better at Oxley. (N. Harrop/MLS Collection)

The End Draws Near – 1963–1965 • 105

Worcester's 7025 *Sudeley Castle* stored and awaiting disposal days after its withdrawal, 20 September 1964. Its nameplates have already been removed. Note the spotter, notebook in hand, emerging from between the stored engines. (MLS Collection)

7029 *Clun* Castle operating out of Southall after the closure of Old Oak Common and before transfer to Gloucester in June, working the last booked regular steam service from Paddington, the 4.15pm to Banbury, near West Wycombe, 18 May 1965. (D.M.C. Hepburne-Scott/Rail Archive Stephenson)

7029 *Clun* *Castle*, watched by crowds lining the platforms, leaves Paddington with the last steam timetabled train from the terminus, the 4.15pm Paddington-Banbury, 11 June 1965. (Colling Turner/Rail Archive Stephenson)

Gloucester's 7034 *Ince Castle*, withdrawn along with 5042 and 7022 in June 1965, leaving 7029 as the sole survivor in BR operation. It is seen here on its last day of service, 13 June 1965, its nameplate already removed for safe keeping. (N. Fields/MLS Collection)

The sole survivor, 7029 *Clun Castle*, at Gloucester shed, 13 June 1965 – an effort had been made to keep it in respectable order for special train working, unlike its three sisters at Gloucester. (N. Fields/MLS Collection)

108 • GREAT WESTERN CASTLE CLASS 4-6-0 LOCOMOTIVES – THE FINAL YEARS 1960–1965

Once one of Old Oak Common's 'star' locomotives, *7036 Taunton Castle*, presents a sorry sight awaiting disposal after its withdrawal in September 1963. It is seen here the following month, 6 October 1963. (MLS Collection)

5027 Farleigh Castle, identified only by the chalked scrawl in place of the numberplate, withdrawn from Llanelli shed in November 1962 only nineteen months after rebuilding with a double chimney. It is stored here over six months later at Tyseley awaiting disposal for scrapping, 30 June 1963. (MLS Collection)

At the end of the 1962 summer timetable, twenty-seven Castles were withdrawn as redundant. Here at Swindon Works awaiting disposal on 30 September is a line of six including, from the camera, 4085 *Berkeley Castle* and 5090 *Neath Abbey*, both of which had been withdrawn earlier that year in May, also 5012 *Berry Pomeroy Castle* (withdrawn April 1962), 5035 *Coity Castle* (withdrawn May 1962), 5024 *Carew Castle* (withdrawn May 1962) and 5006 *Tregenna Castle* (withdrawn April 1962). Three more can be seen in the shed beyond. They have all had their brass safety valve covers removed but still retain brass number and nameplates. (R.O. Tuck/Rail Archive Stephenson)

Chapter 5
CONCLUSIONS

The Great Western Railway and British Railways (Western Region) relied on the Castles for forty years for the successful haulage of its express services to London, Plymouth and Penzance, Bristol, Swansea and Fishguard, Gloucester, Worcester and Hereford, Wolverhampton, Shrewsbury and Chester. It built the first one in 1923 and withdrew the first Castle (a rebuilt 'Star' whose frames were forty-three years old) while still constructing the last ten of the class in 1950. The early exploits of 4074, 4076, 4079 and 5000 on GWR, LNER and LMS metals exerted a significant influence over the design of express locomotives of the other railway companies. In the early days they demonstrated economy whilst hauling huge loads to Taunton, Exeter and Plymouth, whilst in the 1930s they showed their brilliance in high speed performance on the *Cheltenham Flyer* and *Bristolian*, culminating in the holding of the world steam speed record for the fastest start-to-stop journey time of 81.7mph of 5006 *Tregenna Castle* in 1932, breaking the previous record of 5000 *Launceston Castle* of 79.6 in September 1931, which in turn had surpassed the previous record of the USA's Reading Railroad of 78.6mph over 55 miles between Camden (Philadelphia) and Atlantic City.

In the late 1950s and early 1960s, many in double-chimney form, they performed on 60mph+ schedules on several routes and beat the best pre-war *Bristolian* times frequently with 7018's fastest run of 93 minutes 50 seconds in July 1958. The engines had a remarkable safety record. They were fitted with the GW ATC system and only appear to have been involved in two train accidents involving fatalities, both in situations where the accident was caused by another train – 4091 colliding head-on with an LMS 8F overrunning its signals at Slough in 1941 and 5009 on the 1.25am Paddington-Swansea sleeper running into wreckage after 4707 overran its loop at Swindon with a freight and derailed in the 1950s.

The Castles suffered a period of difficulty in the immediate post-war years when there were huge arrears of maintenance and the availability of good steam coal for which they were designed was in very short supply. Despite Hawksworth designing his 'County' class in 1945, it was more Castles that the GWR Board and then the BTC authorised between 1946 and 1950, with the small increase in superheating capacity to cope better with the fuel quality. Their resurgence was significant enough for the Western Region authorities to introduce a host of major train accelerations in the 1954 summer timetable, including the restoration of the pre-war *Bristolian* and additional high speed services like the *Pembroke Coast Express* and the *Cambrian Coast Express*. Whilst a couple of these commenced operation with 'Kings', it was soon realised that Castles could handle these equally well.

The BR Swindon engineers then developed the Castle 4-row superheater boiler, experimented with improved draughting and finally equipped sixty-six of the class with those boilers and double chimneys making them as effective in post-war years as they had been in their prime in the 1930s. Many train schedules did not demand high output, of course, and the Castles were economical engines on the road, and also achieved consistently high mileage between Works repairs significantly better than the equivalent locomotives of similar power on the other Regions.

In the 1950s and 1960s, the Castles seemed most at home with loads up to ten or eleven coaches, with 'Kings' on the heavier trains. Many WR trains, especially on the South Wales route (apart from the *Pembroke Coast Express* and the *South Wales Pullman*), regularly ran to twelve or thirteen coaches and here the Castles needed to be worked on full regulator with 17-18 per cent cut-off and not the 50 per cent (1st port) regular opening that many WR drivers used on the less demanding services, which would hold Castles with these loads at 60-68mph on level track. One of the reasons for the success of the Castles in my opinion was the competence and diligence of Western firemen, all of whom seemed thoroughly used to 'little and often' as their firing technique. I had thirty-one runs on the footplate of Castles and on none of them did I see any signs of shortage of steam, or indeed any concern at all. On the only run where there was a temporary lapse, it was established within ten minutes that the cause was the failure to close the smokebox door completely on shed and once this was put right, steaming was rock steady on 225lb psi for the rest of the journey.

Between 1952 and 1964 I had 549 runs behind Castles, 106 of the 1923-27 '4073-5012' batch, 294 behind the 1932-39 '5013-97' series and 149 behind the Hawksworth '1946-50' built engines. Of these 132 were longer distance runs, the others being under fifty miles on runs like Paddington-Reading, Reading-Oxford, Swansea-Cardiff and Swansea-Carmarthen. In that whole time, I only experienced two occasions when the Castle had to be replaced en route – one was the famous time when 4079 melted its firebars whilst travelling at 96mph on the *Great Western* high speed run of May 1964 and the other when 5037 *Monmouth Castle*, itself a substitute for a failed 'Hymek', ran splendidly from Swansea to Cardiff and there suffered a burst steam heating pipe between engine and train. The Cardiff station staff couldn't find a spare, so they replaced 5037 with their standby 5096 which unfortunately was in poor condition and wasn't able to recover the time lost at Cardiff (they'd have done better just to pinch the steam heating pipe off 5096 rather than change the engine!).

I can trace just five occasions when Castles lost time in running due to defects, apart from the three Saturday *Pembroke Coast Expresses* when they lost time by the public timetable but kept time by the revised timings issued that day and when the engines were clearly in excellent condition and their drivers chose to take it easy rather than attempt to keep to the public book. Four were caused by steaming problems – 4094 had blocked tubes on the Up *Pembroke Coast Express* when it was a last minute replacement for double chimney 4097 which had been held back by Landore to cover the *South Wales Pullman*, 5066 on the Up Oxford 'Flyer' after the fire had become badly clinkered on Oxford shed when left unattended before the run, 5045 on the last Up Wolverhampton when there was a slight loss of time debitable to the engine and I could hear the blower on for part of the way and a run from Reading to Paddington where 5093 was struggling with blocked tubes on a heavy Up South Wales express. The other occasion was in the last throes of 7034's existence when it struggled with a heavy North & West express substituting for a 'Warship' with steam oozing from every pore and had to stop for a banker at Abergavenny.

In all I recorded or noted outline running of 4,646 steam journeys, of which 772 were longer distance (50 miles +) runs, 267 on the Western of which in addition to the 132 Castle runs 67 were behind 'Kings', 177 on the Eastern, 144 on the Southern, 128 on the London Midland and 56 behind BR Standard locomotives. The only locomotives behind which I experienced a personal 100mph+ were three Castle maxima. I had seven occasions when Castles hit between 94 and 98mph. My next highest speed was one rebuilt 'Merchant Navy' that touched 97, a 'King' at 96, three rebuilt 'West Countries' at 95, one LM 'Duchess' at 94, an 'A4' at 94 and three double chimney 'A3s' at 93. I had no 'Royal Scot', 'Patriot', 'Jubilee', 'A1', 'A2', 'Britannia' or 'Lord Nelson' as high as 90mph. The only other class of locomotive that I timed at 90mph was an LM 'Princess Royal' in the descent of Beattock.

When power categories were first established by the BR motive power authorities, they initially classified the Castles as '6P' based on their tractive effort and size. In my view, no other engine has so consistently 'punched above its weight' and based on the work the Castles were expected to perform and did, their power classification was raised to '7P'. One striking aspect of the Castles – like other

Great Western engines – was their adhesion and therefore their ability to start, accelerate and climb with the minimum of wheel slip. The fast starts of which they were capable outshone the locomotives of the other UK railways. The Great Western Railway was always a past master at publicity and boosted their 'Kings and Castles' in a way that made them favourites of many and gained them more fans than those of any other railways' locomotives (and also more jealousy/envy?). Not only did they perform well, they looked good. A Castle in all its splendour, with polished copper and brass work and lovely deep green had, in my humble opinion, no equal. They were many people's favourite engines and they were mine.

For sheer spectacle, I vividly remember standing on Reading platform waiting for a London train. The main line signals on the centre road were off, and there was billowing white smoke on the horizon, then suddenly a piercing shriek as the train reached the junction at the west end of the station and 4085 *Berkeley Castle*, gleaming in the sun, pounded through the station at full throttle, the exhaust a machine gun roar – it must have been doing 80mph – followed by the immaculate chocolate and cream coaches of the *Cheltenham Spa Express*. Every eye looked up and followed the sight until the tail light disappeared rapidly towards Sonning. Just one memory beats that, and that was being on the fireman's seat of 7030 *Cranbrook Castle* as we rocketed through Reading on the evening 'Oxford Flyer' at nearly 85 mph!

COLOUR SECTION

Castles in the South West

5058 *Earl of Clancarty* passing its home depot, Plymouth Laira, with the Up *Cornishman* to Wolverhampton, June 1960. (Brian Penney)

4077 *Chepstow* Castle with the 3.5pm Exeter-Kingswear stopping train at Churston, June 1960. The Brixham branch is off to the right. (Peter Gray/GW Trust)

4087 *Cardigan* Castle leans to the curve descending Dainton Bank and approaching Brent with the 10.29am Manchester-Plymouth, 2 August 1960. (Peter Gray/GW Trust)

5066 *Sir Felix Pole* pilots 5065 *Newport Castle* on the 7.30am Penzance-Manchester express over the South Devon banks descending from Dainton past Stoneycombe Quarry, 3 September 1960. (Peter Gray/GW Trust)

4081 *Warwick Castle* accelerates away from Aller Junction with a Sunday express from Manchester for Plymouth, 7 February 1960. (Peter Gray/GW Trust)

116 • GREAT WESTERN CASTLE CLASS 4-6-0 LOCOMOTIVES – THE FINAL YEARS 1960–1965

Newton Abbot's 4037 *The South Wales Borderers* braves the sea water approaching Dawlish station with the double-home Shrewsbury – Newton Abbot diagram on the 9.5am Liverpool-Plymouth, 17 September 1960. This engine was Newton Abbot's regular engine on this turn until April 1961. (Peter Gray)

4037 *The South Wales Borderers*, recently returned from its last Swindon overhaul, on a Kingswear-Swansea train at Burlescombe on the climb to Whiteball Tunnel, 3 September 1960. (K.L. Cook/Rail Archive Stephenson)

5053 *Earl* Cairns of Newton Abbot climbs Dainton Bank with the return 6.10pm Goodrington-Plymouth excursion, summer 1960. (Peter Gray)

4087 *Cardigan* Castle at Yatton with an Up special train from Plymouth, 24 March 1962. (Peter Fry/GW Trust)

Stafford Road's 5022 *Wigmore Castle* departs from Exeter St David's with the down *Cornishman*, 26 August 1961. (Peter Gray/GW Trust)

Laira's 5058 *Earl of Clancarty* with a Cardiff - Newquay express at Liskeard, 15 July 1961. (Peter Gray/GW Trust)

Old Oak Common's 7036 *Taunton Castle* with a Summer Saturday Paddington-Paignton express in Sonning Cutting, 9 September 1961. (Derek Penney)

Laira's 4087 *Cardigan Castle* departing from Bristol with the Sunday 1.30pm Bristol Temple Meads-Plymouth express, 25 March 1962. (Peter Gray/GW Trust)

5051 *Earl* Bathurst at Whiteball summit with the 12.18pm holiday relief from Newton Abbot for Paddington, 1 July 1961. 5051 appears to be ex-works and still running in before being returned to its Landore depot. (Peter Gray/GW Trust)

5003 *Lulworth* Castle passing non-stop through Newton Abbot station with the Down *Torbay Express* to Kingswear, 25 June 1961. (Peter Gray/GW Trust)

Newton Abbot's 5055 *Earl of Eldon* descending from Whiteball Tunnel with a Kingswear-Paddington express, 9 June 1962. (Peter Gray/GW Trust)

5092 *Tresco Abbey* brings a heavy 13-coach train from Perrenporth and Falmouth for Paddington into Taunton station, with small prairie tank 5563 in the background, 11 August 1962. (Peter Gray/GW Trust)

7017 *G.J. Churchward* awaits departure with the 12.18pm to Exeter stopping train while Hymek D7045 pulls ahead westwards from Taunton station, 11 August 1962. BR Standard 3 tank 82030 is in the sidings. (Peter Gray/GW Trust)

Reading's 4096 *Highclere Castle* pauses at signals outside Exeter St David's with an Up empty stock train while the fireman takes advantage to pull some coal forward, October 1962. (Bruce Oliver)

Stafford Road's 5089 *Westminster Abbey* approaching Exeter St David's with a Paignton-Manchester Summer Saturday express, 15 June 1962. (Bruce Oliver)

5089 *Westminster Abbey* accelerating away from Newton Abbot at Hackney with a train for Paddington, June 1962. (GW Trust)

One of Laira's last two working Castles kept as a standby for diesel failures, 7022 *Hereford Castle,* climbing Dainton Bank with a Plymouth-Manchester express, summer 1962. (Peter Gray)

Laira's 4087 *Cardigan Castle* with new front end section, double-chimney, 4-row superheat boiler and valveless mechanical lubricator at its home depot, 27 January 1962. (Peter Gray/GW Trust)

Castles in the London area

7022 *Hereford* Castle again, coming off the Berks & Hants line at Reading with the 1.28pm Saturday Paignton-Paddington express, 8 September 1962. Note the spotters sitting on the platform edge. (Derek Penney)

5011 *Tintagel* Castle, a Newton Abbot engine for many years but ending its days at Old Oak Common, passing through Tilehurst with the 10.35am Weston-super-Mare-Paddington, 1 September 1962. (Derek Penney)

Bristol St Philip's Marsh's 5052 *Earl of Radnor* passes through Tilehurst with the 10.35am from Weston-super-Mare in September 1962, a few days before withdrawal. St Philip's Marsh had few cleaners and 5052 was distinctly grubby though seemingly steam tight. (Derek Penney)

Neath's 5075 *Wellington*, with Landore depot's landmark silver painted buffers still visible, heads the 11.55am Paddington-Milford Haven through Sonning Cutting, 9 September 1961. (Derek Penney)

Canton's double chimney 5097 *Sarum Castle* heads the 3.55pm Paddington *Capitals United Express* past Tilehurst, 1961. (Derek Penney)

Landore's 4076 *Carmarthen Castle* passing West Ealing with the 11.55am Paddington–West Wales, 9 May 1962. (Charles Gordon-Stuart/ GW Trust)

Landore's 5078 *Beaufort* passing West Ealing with the down *Pembroke Coast Express*, 16 May 1962. (Charles Gordon-Stuart/GW Trust)

Old Oak Common's 5034 *Corfe Castle* passing West Ealing with the 11.55am Paddington-West Wales express, 24 February 1962. (Charles Gordon-Stuart/GW Trust)

Colour Section • 129

7000 *Viscount* Portal passing West Ealing with the Up *Cheltenham Spa Express,* 16 March 1962. (Charles Gordon-Stuart/ GW Trust)

Gloucester's 7034 *Ince Castle* with the 11.45am Cheltenham that combined with the 11.40am Weston-super-Mare at Swindon, near Pangbourne, May 1962. It bears the train reporting number of a Down train, probably its previous working. (Derek Penney)

Castles in South Wales

Canton's 5081 *Lockheed Hudson* enters Cardiff General station with a Plymouth-Swansea train, passing Collett 2-8-2T 7205 on an Up oil train, April 1961. (Alan Jarvis/SLS)

Landore's 7028 *Cadbury Castle* waits to depart from Cardiff General with the 2.30pm Neyland-Paddington as a Canton 'King' arrives with the Down *Capitals United Express*, June 1961. The excellent external condition of 7028 is remarkable as it is only two months before a Swindon Heavy General overhaul when it would receive a double chimney. (Alan Jarvis/SLS)

Old Oak Common's 5093 *Upton Castle* passing Leckwith Junction with a West Wales-Paddington express, August 1961. (Alan Jarvis/SLS)

Landore's 4090 *Dorchester Castle* stands at Swansea High Street ready to depart with the last steam-hauled *South Wales Pullman* before the introduction of the 'Blue Pullman' diesel unit, 15 September 1961. (Alan Jarvis/SLS)

4090 *Dorchester* Castle with the last Up *South Wales Pullman* at Cardiff General in the evening sun of 15 September 1961. (Alan Jarvis/SLS)

Old Oak Common's 5015 *Kingswear Castle* at Cardiff General on the 8am Neyland-Paddington, September 1961. (Alan Jarvis/SLS)

Reading's 5018
St Mawes Castle at Leckwith Junction accelerating from the Cardiff General stop, with the 11.55am Paddington-West Wales, on Boxing Day, 26 December 1961.
(Alan Jarvis/SLS)

4076 Carmarthen
Castle passing Cardiff Goods Shed and Canton Sidings with the 8.55am Paddington-West Wales, 18 February 1962.
(Alan Jarvis/SLS)

Landore's 5051 *Earl Bathurst* passes Cardiff Goods Shed and Canton Sidings on a crisp winter's day with the 10.55am Paddington *Pembroke Coast Express*, 6 March 1962. (Alan Jarvis/SLS)

Landore's double chimneyed 4090 *Dorchester Castle* a week later at the same location, just west of Cardiff General station with the Down *Pembroke Coast Express,* 13 March 1962. (Alan Jarvis/SLS)

Landore's 5037 *Monmouth Castle* drifts past Ely Main Line station with the 11.50am Swansea-Manchester express, 4 June 1962. (Alan Jarvis/SLS)

Landore's double-chimney 4093 *Dunster Castle* passes onlookers on Ely Main Line station platform with the 11.50am Swansea-Manchester express, 6 June 1962. (Alan Jarvis/SLS)

Landore's 5013 *Abergavenny Castle* approaching Ely with the 11.50am Swansea-Manchester train, 13 June 1962. (Alan Jarvis/SLS)

Canton's 5091 *Cleeve Abbey* heads the Up *Capitals United Express* at St Fagan's, 4 August 1962. It will be booked for a 'Hymek' diesel hydraulic from Cardiff. (Alan Jarvis/SLS)

Llanelli's 5062 *Earl of Shaftesbury*, returned to traffic after storage at Neath, passes St Fagan's with the 9.25am Manchester-Swansea, 4 August 1962.
(Alan Jarvis/SLS)

Gloucester's 5064 *Bishop's Castle* with the 11.35am Paddington-Swansea at St George's, 18 August 1962.
(Alan Jarvis/SLS)

Carmarthen's 5054 *Earl of Ducie* has backed on to the Down *Pembroke Coast Express* at Swansea High Street to haul it to Carmarthen where it will reverse again, 26 April 1963. GW pannier tank 9752 is getting ready to attach to assist the train to Cockett Tunnel.
(Alan Jarvis/SLS)

5091 *Cleeve* Abbey with the empty stock for a special train, Z43, at Cardiff General station, May 1963. (Michael Hale/GW Trust)

Canton's 5048 *Earl of Devon* at its home depot, March 1961. It has a 4-row superheater and would be withdrawn from Llanelli shed in August 1962. (Alan Jarvis/SLS)

5021 *Whittington Castle* at its home depot of Canton with a pannier tank and in the background a 9F 2-10-0, a WD 2-8-0 and even an LMS 4F, April 1961. (Alan Jarvis/SLS)

5043 *Earl* of Mount *Edgcumbe* and **4080 *Powderham Castle*** on Canton shed with **6847 *Tidmarsh Grange***, 23 July 1962. (Alan Jarvis/SLS)

4080 *Powderham* Castle at Canton, a former regular engine for the Up *Red Dragon* and Down *Capitals United Express*, now sidelined by the new 'Hymeks' allocated to Landore, 17 August 1962. It achieved the highest mileage of any of the Castles other than a few of the rebuilt 'Stars'. It was not withdrawn until August 1964.
(Alan Jarvis/SLS)

Castles on the Northern routes

Worcester's 7004 *Eastnor Castle* passes through Lapworth station with a Birmingham-bound express, which from the mixed rolling stock behind it, is probably a cross country train from the South Coast via Reading or Weymouth via Swindon, August 1961. (Derek Penney)

Oxford's 5038 *Morlais Castle* climbing Hatton bank with a Southern Region-West Midlands express, 7 July 1962. (Michael Hale/GW Trust)

4090 *Dorchester Castle* with the 2.3pm Chester-Wolverhampton train at Gobowen station, 23 August 1962. The Gobowen – Oswestry auto train (with a 14XX 0-4-2T) is in the bay. (Peter Gray/GW Trust)

Reading's 4074 *Caldicot Castle* climbing Hatton bank with little apparent effort on the 9.25am Margate-Wolverhampton relief to the 'Conti', August 1962. Although 4074 has received a 4-row superheater boiler and double chimney, it still has the original 'joggled' frame (note the narrow inside cylinder block casing).
(Derek Penney)

4096 *Highclere* Castle, recently transferred from Old Oak Common to Reading, with the 10.42am Wolverhampton-Ramsgate formed of Southern Region stock, at Hatton, Summer 1962.
(Derek Penney)

A decent attempt to clean up 7008 *Swansea Castle* by the Oxford cleaning gang is evident as it powers the 9.20am Bournemouth-Wolverhampton up Hatton bank, Summer 1962. (Derek Penney)

Old Oak Common's 7018 *Drysllwyn Castle* is working hard on the climb of Hatton bank with a Fulham football club supporters' special train to an FA Cup semi-final match at Villa Park, 31 March 1962. (Derek Penney)

7001 *Sir* James Milne at Wolverhampton with the southbound *Cornishman*, 7 September 1962.
(Michael Hale/GW Trust)

Castles on the North & West

5043 *Earl* of Mount Edgcumbe passes Stokesay Castle near Craven Arms with the 12.28pm Manchester–Cardiff, 19 August 1961. 5043 was recorded at Old Oak Common at this time and had presumably been borrowed by Canton before its transfer to Cardiff East Dock a few months later.
(Derek Penney)

Shrewsbury's 5059 *Earl St Aldwyn* on the 8am Plymouth-Liverpool near Craven Arms, 19 August 1961. (Derek Penney)

7015 *Carn Brea Castle* is standby at Shrewsbury depot for a Manchester-Plymouth train booked for haulage by D800 *Sir Brian Robertson* which has worked north on the 8am Plymouth-Liverpool, March 1962. (Brian Penney)

Castles on the Gloucester and Worcester routes

Gloucester's 7000 *Viscount Portal* descends from Sapperton through Brimscombe with the Gloucester/Cheltenham portion of a Paddington-Bristol express, c1962. (Peter Gray/GW Trust)

Worcester's 7004 *Eastnor Castle* at Fladbury with the 11.15am Paddington-Hereford during the summer of 1962. A number of these photographs show the reporting name of the previous working still displayed. (Derek Penney)

Worcester's 7007 *Great Western* in superb condition with the Up *Cathedrals Express* near Ascott-under-Wychwood, May 1962. I rode from Reading to Paddington on the footplate of this engine the same month and can confirm that it was in excellent mechanical condition and accelerated to 75-80mph by Maidenhead with full regulator and 17 per cent cut-off. The beat was sharp, even and a joy to hear. (Derek Penney)

Another engine I footplated in late April 1962, this time from Oxford to Paddington, 7013 *Bristol Castle*, the 1924 built 4082 *Windsor Castle*, seen here on the 7.35am Worcester near Ascott-under-Wychwood, which will return on the 11.15am Paddington, May 1962. Although though not as well polished as 7007 and other Worcester Castles, it was still in excellent condition and rode smoothly with plenty of steam. (Derek Penney)

Worcester's 7031 *Cromwell's Castle,* a Laira engine until the beginning of 1960, is photographed at Pangbourne on the 1.15pm from Paddington on 28 April 1962 on which I was enjoying my first footplate trip down to Oxford with my newly acquired cab pass during my three months' training at Old Oak Common. (Derek Penney)

7007 *Great* Western passing West Ealing with the Up *Cathedrals Express*, 9 May 1962. (Charles Gordon-Stuart/ GW Trust)

Colour Section • 151

7023 *Penrice* Castle brings the empty stock of a Paddington express into Hereford station, Summer 1962. A Castle which had arrived from London drew the stock out beyond the bridge and 7023 then backed on to head the train back to London. (Bruce Oliver)

An engine that spent its entire career at Worcester, 7005 *Sir Edward Elgar* (*Lamphey Castle* until August 1957) draws into Evesham station with an Up Worcester express in the Spring of 1963. (Derek Penney)

Worcester's 7023 *Penrice Castle* climbing Honeybourne bank to Chipping Campden Tunnel with a morning Worcester-Paddington train in the last summer of booked steam working of the Worcester trains, 1963. 7023 was transferred, like a number of the Hawksworth Castles, to Worcester in 1960 and it would move to Oxley shed in 1964. (M. Mensing)

7027 *Thornbury* Castle at Chipping Camden with a Worcester-Paddington express, July 1963. Cleaning standards at Worcester have slipped in their last summer of booked steam on the London expresses. (Michael Hale/GW Trust)

7023 *Penrice Castle* poses at Worcester depot after working the last diagrammed express steam train between Paddington and Worcester, 9 September 1963. In fact, because of shortages and failures of the replacement 'Hymeks', 7023 and its Worcester sisters appeared on many occasions right through to the summer of 1964, when 7023 moved to Oxley. (Derek Penney)

Castles on Shed or Works

Two of the last Castles to receive a full overhaul at Swindon Works stand ready to be paired with repainted tenders, 5043 *Earl of Mount Edgcumbe* and 7026 *Tenby Castle*, April 1962. (Derek Penney)

7004 *Eastnor Castle* undergoes a heavy repair alongside mogul 7337, 5037 *Monmouth Castle* and 1009 *County of Carmarthen* in Swindon Works, 8 April 1962.
(Peter Gray/GW Trust)

Bristol St Philip's Marsh's 5071 *Spitfire*, after a casual repair at Swindon Works, at Stafford Road shed, 11 May 1963.
(Michael Hale/GW Trust)

Castles on freight and parcels work

7034 *Ince Castle* at West Ealing with a Down milk empties train, 14 February 1962. (Charles Gordon-Stuart/ GW Trust)

Old Oak Common's *7017 G.J. Churchward* leads a parcels train on the Up relief line between Sonning and Twyford, 10 September 1960. (Ken Wightman)

Old Oak Common's double chimney 5032 *Usk Castle* heads a long Up parcels train near Pangbourne in September 1962, still in external good condition, but shortly to be redundant and withdrawn at the end of the month. (Derek Penney)

7011 *Banbury* Castle passes Berkeley Road with a fitted freight, 16 May 1964. (Charles Gordon-Stuart/GW Trust)

5057 *Earl* Waldegrave enters Birmingham Snow Hill with a parcels train from Shrewsbury, 1963. (Michael Hale/GW Trust)

Castles on Special workings

5050 *Earl* of St Germans of St Philip's Marsh was impounded at Fratton depot after working a special train to Portsmouth in June 1963. It is seen here awaiting repatriation alongside a couple of Fratton's 'Terriers' used on the Hayling Island branch, 32650 and 32646. (Bruce Oliver)

7032 Denbigh Castle passing West Ealing with a Newbury race special, 27 March 1963. (Charles Gordon-Stuart/GW Trust)

7029 Clun Castle passing West Ealing with a Newbury race special, 14 March 1963. This engine was involved in the very last steam hauled Newbury Race Specials. (Charles Gordon-Stuart/GW Trust)

9 May 1964 – The 'Great Western'

The lucky passengers swarm round the engine and crew of the first leg of the 9 May 1964 'Castle Swan Song', the Ian Allan *Great Western*, with 4079 *Pendennis Castle* in platform 2 at Paddington before departure. The author is seen with hands in pocket next to the smokebox of the second photo.
(Bruce Oliver)

4079 *Pendennis* Castle is a picture as it accelerates away from the Reading curve slack towards Southcote Junction with the *Great Western* special, 9 May 1964.
(Derek Penney)

7025 *Sudeley* Castle with the *Great Western* special at Plymouth North Road on arrival, after having replaced Westbury's temporary substitute 6999 for the disabled 4079 at Taunton, 9 May 1964.
(Bruce Oliver)

4079 is stripped of its headboard and reporting number as it has to be removed from the *Great Western* special at Westbury after dropping melting firebars while travelling at 96mph after Lavington, 9 May 1964. I am glimpsed once more on the opposite platform, the young man with camera and 'flying' tie! (Bruce Oliver)

Plymouth Laira's standby engine for the *Great Western* special, 7008 *Swansea Castle*, draws the empty stock into Plymouth North Road which 7029 *Clun Castle* will take over when it is fully prepared, 9 May 1964. (Bruce Oliver)

Bill Rundle was one of two firemen on the footplate of 7029 *Clun Castle* from Plymouth to Bristol on that 1964 occasion. He is seen here with his wife, Joyce, on the GW 175th Anniversary *Bristolian* run in April 2010. Unfortunately he died just a couple of months before the 50th commemorative run of the May 1964 record run, but his brother was taken on 5043's footplate and put Bill's ashes in the fire as they ascended Hemerdon Bank. (Bob Meanley's Collection)

Castles in the last years

By the autumn of 1963 'Hymeks' should have been in charge of the Worcester line trains but steam power was used to cover diesel failures or shortages well into the summer of 1964. Worcester's 7005 *Sir Edward Elgar,* suffering from lack of attention since dieselisation, covers a Down Worcester express, probably the 1.15pm Paddington, at Pangbourne, 24 August 1963. (Bruce Oliver)

The last booked Worcester steam turn to Paddington with 7023 *Penrice Castle* on the 11.10am departure from Worcester Shrub Hill, 7 September 1963. (T. Done/John Scott-Morgan Collection)

5039 *Rhuddlan* Castle had operated out of Goodwick (Fishguard) during the summer of 1963 and had spent the autumn in store, being brought back into traffic at the beginning of 1964 and allocated to Cardiff East Dock. It is arriving at Reading with a westbound express, 14 April 1964 – the result of a diesel failure? (Bruce Oliver)

5039 *Rhuddlan* Castle waits to depart from Reading. The man on the platform is holding what looks like operating notices – perhaps this was an Easter relief train, 14 April 1964. (Bruce Oliver)

Cardiff East Dock's 5091 *Cleeve Abbey* was transferred to Worcester on 13 April 1964 and just one day later, smartened up by Worcester shed staff, 5091 is already in action covering a Worcester-Paddington diesel turn, seen stopping at Reading, 14 April 1964. (Bruce Oliver)

Worcester Hymeks are clearly having a bad day on 14 April 1964 as we see another of Worcester's dwindling number of Castles in action, 7025 *Sudeley Castle* on a Down service stopping at Reading, 14 April 1964. (Bruce Oliver)

5063 *Earl* Baldwin at Honeybourne with a West of England-Wolverhampton return holiday express, (1M35), 4 July 1964. (Michael Hale/GW Trust)

5056 *Earl of Powis* of Oxley depot approaches Oxford with a Wolverhampton-Margate train of Southern Region stock, 4 July 1964. (Charles Gordon-Stuart/GW Trust)

7019 *Fowey* Castle of Oxley at Yate with a southbound express from Wolverhampton to Penzance, 22 August 1964. (Charles Gordon-Stuart/GW Trust)

7029 *Clun* Castle with the last scheduled steam working from Paddington, the 4.15pm to Banbury, passing North Acton, 11 June 1965. (Charles Gordon-Stuart/GW Trust)

7029 *Clun Castle* with the 'Farewell to Steam' special to Bristol and back to Swindon, passing Kensal Green, 27 November 1965. (Colour Rail)

APPENDIX

Dimensions

Cylinders (4)	16in x 26in
Coupled wheel diameter	6ft 8½ in
Bogie wheel diameter	3ft 2in
Boiler pressure	225lb psi
Heating surface	2,280.7sqft (2,258sqft 5098 class with 3-row superheat)
Grate area	29.36sqft
Axleload	19¾ tons
Weight	
- Engine	79 tons 17 cwt
- Tender	46 tons 14 cwt
- Total	126 tons 11 cwt
Water capacity	4,000 gallons
Coal capacity	6 tons
Tractive effort	31,625lb

Weight diagram

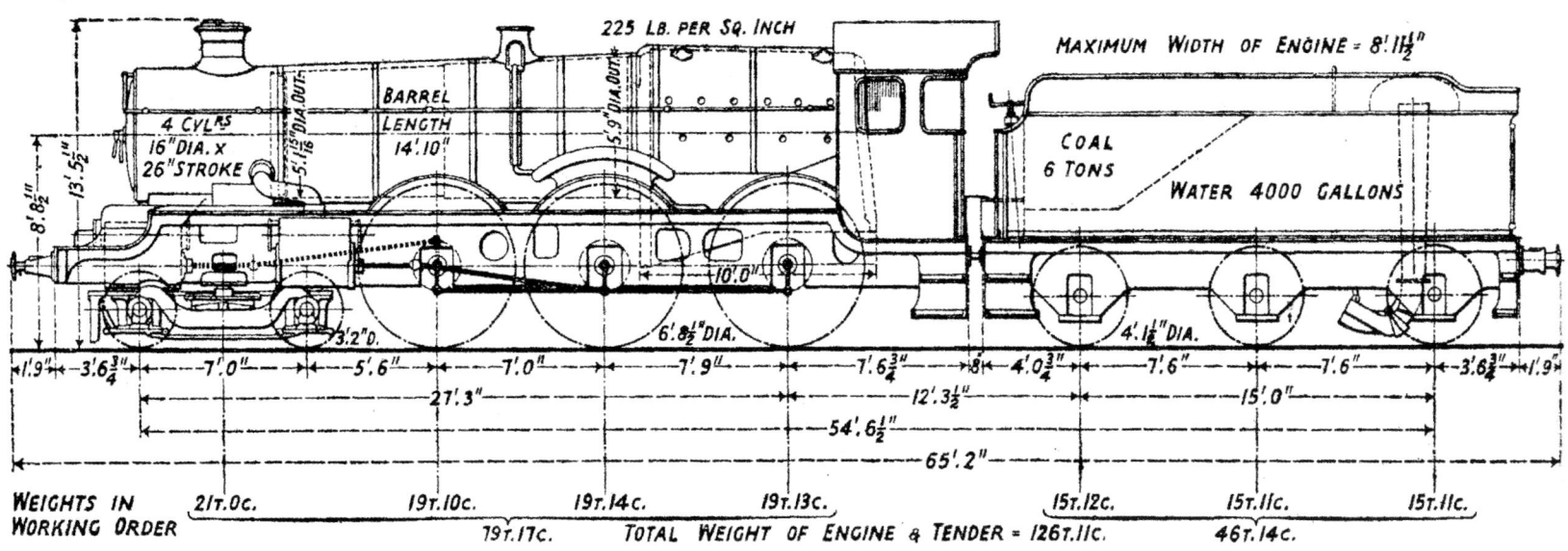

Statistics

No.	Built	Name	Double chimney	Withdrawn	Mileage
111	9/24	*Viscount Churchill*		7/53	1,989,628**
		(was *The Great Bear* from 1908)			
4000	11/29	*North Star*		5/57	2,110,396**
4009	4/25	*Shooting Star* 1/36 100A1 *Lloyd's*		3/50	1,974,651**
4016	10/25	*Knight of the Golden Fleece*			
	2/38	*The Somerset Light Infantry (Prince Albert's)*		9/51	1,972,559**
4032	10/25	*Queen Alexandra*		9/51	1,981,335**
4037	6/26	*Queen Philippa*		9/62	2,429,722**
	4/37	*The South Wales Borderers*			
4073	8/23	*Caerphilly Castle*		5/60 Preserved	1,910,730
4074	12/23	*Caldicot Castle*	4/59	5/63	1,844,072
4075	1/24	*Cardiff Castle*		11/61	1,807,802
4076	2/24	*Carmarthen Castle*		2/63	1,697,895
4077	2/24	*Chepstow Castle*		8/62	1,823,488
4078	2/24	*Pembroke Castle*		7/62	1,917,380
4079	3/24	*Pendennis Castle*		5/64 Preserved	1,758,398
4080	3/24	*Powderham Castle*	8/58	8/64	1,974,461
4081	3/24	*Warwick Castle*		2/63	1,894,998
4082	4/24	*Windsor Castle*	5/58	2/65	1,898,571
	7013	*Bristol Castle* from 2/52			
4083	5/25	*Abbotsbury Castle*		12/61	1,677,060
4084	5/25	*Aberystwyth Castle*		10/60	1,674,812
4085	6/25	*Berkeley Castle*		5/62	1,651,000
4086	6/25	*Builth Castle*		4/62	1,791,633
4087	6/25	*Cardigan Castle*	2/58	10/63	1,812,341
4088	7/25	*Dartmouth Castle*	5/58	5/64	1,848,430*
4089	7/25	*Donnington Castle*		9/64	1,876,807
4090	8/25	*Dorchester Castle*	4/57	6/63	1,848,646
4091	8/25	*Dudley Castle*		1/59	1,691,856
4092	8/25	*Dunraven Castle*		12/61	1,718,879
4093	5/26	*Dunster Castle*	12/57	9/64	1,842,985*
4094	5/26	*Dynevor Castle*		3/62	1,881,886
4095	6/26	*Harlech Castle*		12/62	1,695,899

No.	Built	Name	Double chimney	Withdrawn	Mileage
4096	6/26	Highclere Castle		2/63	1,958,378
4097	6/26	Kenilworth Castle	6/58	12/60	1,713,966
4098	7/26	Kidwelly Castle		12/63	1,723,879
4099	8/26	Kilgerran Castle		9/62	1,873,985
5000	8/26	Launceston Castle		10/64	1,870,200*
5001	8/26	Llandovery Castle	7/61	2/63	1,885,495
5002	8/26	Ludlow Castle		9/64	1,817,218
5003	5/27	Lulworth Castle		8/62	1,698,751
5004	6/27	Llanstephan Castle		4/62	1,854,704
5005	6/27	Manorbier Castle		2/60	1,731,868
5006	6/27	Tregenna Castle		4/62	1,812,966
5007	6/27	Rougemont Castle		9/62	1,854,951
5008	6/27	Raglan Castle	3/61	9/62	1,798,646
5009	7/27	Shrewsbury Castle		10/60	1,708,246
5010	7/27	Restormel Castle		10/59	1,684,146
5011	7/27	Tintagel Castle		9/62	1,732,565
5012	7/27	Berry Pomeroy Castle		4/62	1,625,965
5013	7/32	Abergavenny Castle		7/62	1,525,662
5014	7/32	Goodrich Castle		2/65	1,615,297*
5015	7/32	Kingswear Castle		4/63	1,554,288
5016	7/32	Montgomery Castle	2/61	9/62	1,480,896
5017	7/32	St Donat's Castle		9/61	1,598,851
	4/54	The Gloucestershire Regiment 28th 61st			
5018	7/32	St Mawes Castle		3/63	1,503,642
5019	7/32	Treago Castle	3/61	9/62	1,521,335
5020	8/32	Trematon Castle		11/62	1,636,749
5021	8/32	Whittington Castle		9/62	1,446,936
5022	8/32	Wigmore Castle	2/59	6/63	1,546,104
5023	4/34	Brecon Castle		2/63	1,479,168
5024	4/34	Carew Castle		5/62	1,351,161
5025	4/34	Chirk Castle		11/63	1,401,530
5026	4/34	Criccieth Castle	10/59	11/64	1,209,457*
5027	5/34	Farleigh Castle	4/61	11/62	1,465,365
5028	5/34	Llantilio Castle		5/60	1,345,291

No.	Built	Name	Double chimney	Withdrawn	Mileage
5029	5/34	Nunney Castle		12/63 Preserved	1,523,415
5030	6/34	Shirburn Castle		9/62	1,413,084
5031	6/34	Totnes Castle	6/59	10/63	1,434,409
5032	6/34	Usk Castle	5/59	9/62	1,288,968
5033	5/35	Broughton Castle	10/60	9/62	1,160,197
5034	5/35	Corfe Castle	2/61	9/62	1,250,714
5035	5/35	Coity Castle		5/62	1,444,261
5036	6/35	Lyonshall Castle	12/60	9/62	1,304,430
5037	6/35	Monmouth Castle		3/64	1,500,851
5038	6/35	Morlais Castle		9/63	1,438,862
5039	6/35	Rhuddlan Castle		6/64	1,380,564
5040	7/35	Stokesay Castle		10/63	1,414,142
5041	7/35	Tiverton Castle		12/63	1,383,804
5042	7/35	Winchester Castle		6/65	1,339,221
5043	3/36	Barbury Castle	5/58	12/63 Preserved	1,400,817
	9/37	Earl of Mount Edgcumbe			
5044	3/36	Beverston Castle		4/62	1,377,644
	9/37	Earl of Dunraven			
5045	4/36	Bridgwater Castle		9/62	1,383,737
	9/37	Earl of Dudley			
5046	4/36	Clifford Castle		9/62	1,358,388
	8/37	Earl Cawdor			
5047	4/36	Compton Castle		9/62	1,225,670
	8/37	Earl of Dartmouth			
5048	5/36	Cranbrook Castle		8/62	1,327,811
	8/37	Earl of Devon			
5049	5/36	Denbigh Castle	9/59	3/63	1,282,965
	8/37	Earl of Plymouth			
5050	5/36	Devizes Castle		9/63	1,135,797
	8/37	Earl of St Germans			
5051	5/36	Dryslwyn Castle		5/63 Preserved	1,316,659
	8/37	Earl Bathurst			
5052	5/36	Eastnor Castle		9/62	1,396,894
	7/37	Earl of Radnor			

No.	Built	Name	Double chimney	Withdrawn	Mileage
5053	6/36	*Bishop's Castle*		7/62	1,293,786
	8/37	*Earl Cairns*			
5054	6/36	*Lamphey Castle*		10/64	1,412,394
	9/37	*Earl of Ducie*			
5055	6/36	*Lydford Castle*		9/64	1,439,975
	9/37	*Earl of Eldon*			
5056	6/36	*Ogmore Castle*	11/60	11/64	1,434,833*
	9/37	*Earl of Powis*			
5057	7/36	*Penrice Castle*	7/58	9/64	1,273,324
	10/37	*Earl Waldegrave*			
5058	5/37	*Newport Castle*		3/64	1,224,735
	9/37	*Earl of Clancarty*			
5059	6/37	*Powis Castle*		6/62	1,054,062
	10/37	*Earl St Aldwyn*			
5060	6/37	*Sarum Castle*	8/61	4/63	1,316,240
	10/37	*Earl of Berkeley*			
5061	6/37	*Sudeley Castle*	9/58	9/62	1,020,412
	10/37	*Earl of Birkenhead*			
5062	6/37	*Tenby Castle*		8/62	1,143,143
	11/37	*Earl of Shaftesbury*			
5063	6/37	*Thornbury Castle*		2/65	1,235,058
	1937	*Earl Baldwin*			
5064	7/37	*Tretower Castle*	9/58	9/62	1,155,986
	9/37	*Bishop's Castle*			
5065	7/37	*Upton Castle*		2/63	1,222,961
	9/37	*Newport Castle*			
5066	7/37	*Wardour Castle*	9/58	9/62	1,339,619
	4/56	*Sir Felix Pole*			
5067	7/37	*St Fagans Castle*		9/62	1,192,663
5068	6/38	*Beverston Castle*	3/61	9/62	1,081,514
5069	6/38	*Isambard Kingdom Brunel*	11/58	2/62	1,217,505
5070	7/38	*Sir Daniel Gooch*		3/64	1,139,354
5071	7/38	*Clifford Castle*	6/59	10/63	1,150,913
	9/40	*Spitfire*			

No.	Built	Name	Double chimney	Withdrawn	Mileage
5072	7/38	Compton Castle		10/62	1,055,942
	11/40	Hurricane			
5073	7/38	Cranbrook Castle	5/59	2/64	995,495
	11/40	Blenheim			
5074	7/38	Denbigh Castle		5/64	1,142,187
	1/41	Hampden			
5075	8/38	Devizes Castle		9/62	1,068,502
	10/40	Wellington			
5076	8/38	Drysllwyn Castle		9/64	1,121,080
	1/41	Gladiator			
5077	8/38	Eastnor Castle		7/62	1,089,166
	10/40	Fairey Battle			
5078	5/39	Lamphey Castle	12/61	11/62	1,038,165
	1/41	Beaufort			
5079	5/39	Lydford Castle		5/60	1,008,175
	1/41	Lysander			
5080	5/39	Ogmore Castle		4/63 Preserved	1,117,030
	1/41	Defiant			
5081	6/39	Penrice Castle		10/63	1,208,003
	1/41	Lockheed Hudson			
5082	6/39	Powis Castle		7/62	1,161,413
	1/41	Swordfish			
5083	11/22	Bath Abbey	Reb.6/39	1/59	1,822,834**
5084	12/22	Reading Abbey	Reb.4/37 10/58	7/62	2,017,118**
5085	12/22	Evesham Abbey	Reb.7/39	2/64	2,112,594**
5086	12/22	Malvern Abbey	Reb.12/37	11/58	1,871,501**
	12/37	Viscount Horne			
5087	1/23	Tintern Abbey	Reb.11/40	8/63	2,029,151**
5088	1/23	Llanthony Abbey	Reb.2/39 6/58	9/62	1,879,955**
5089	1/23	Margam Abbey	Reb.11/39	11/64	2,097,247**
	11/39	Westminster Abbey			
5090	2/23	Neath Abbey	Reb.4/39	5/62	2,058,275**
5091	2/23	Cleeve Abbey	Reb.12/38	10/64	1,921,723**
5092	2/23	Tresco Abbey	Reb.4/38 10/61	7/63	1,968,877**

Appendix • 175

No.	Built	Name	Double chimney	Withdrawn	Mileage
5093	6/39	Upton Castle		9/63	1,145,221
5094	6/39	Tretower Castle	6/60	9/62	948,540
5095	7/39	Barbury Castle	11/58	8/62	1,122,493
5096	7/39	Bridgwater Castle		6/64	1,103,607
5097	7/39	Sarum Castle	6/60	3/63	993,804
5098	5/46	Clifford Castle	1/59	6/64	826,525
5099	5/46	Compton Castle		2/63	863,411
7000	5/46	Viscount Portal		12/63	824,873
7001	5/46	Denbigh Castle	9/60	9/63	838,604
	2/48	Sir James Milne			
7002	6/46	Devizes Castle	7/61	3/64	837,626
7003	6/46	Elmley Castle	7/60	8/64	773,642
7004	6/46	Eastnor Castle	2/58	8/64	876,349
7005	6/46	Lamphey Castle		9/64	869,370
	8/57	Sir Edward Elgar			
7006	6/46	Lydford Castle	6/60	12/63	789,052
7007	7/46	Ogmore Castle	6/61	2/63	851,649
	1/48	Great Western			
7008	5/48	Swansea Castle	6/59	9/64	483,663*
7009	7/48	Athelney Castle		3/63	671,920
7010	7/48	Avondale Castle	10/60	1/64	662,192
7011	7/48	Banbury Castle		2/65	748,635
7012	8/48	Barry Castle		11/64	667,408*
7013	7/48	Bristol Castle		9/64	712,286*
	4082	Windsor Castle from 2/52			
7014	7/48	Caerhays Castle	2/59	2/65	765,282*
7015	9/48	Carn Brea Castle	5/59	4/63	636,439
7016	9/48	Chester Castle		11/62	672,533
7017	9/48	G.J. Churchward		2/63	724,589
7018	6/49	Drysllwyn Castle	5/56	9/63	614,259
7019	6/49	Fowey Castle	6/58	2/65	680,454*
7020	6/49	Gloucester Castle	2/61	9/64	610,143*
7021	6/49	Haverfordwest Castle	11/61	9/63	673,241
7022	6/49	Hereford Castle	12/61	6/65	733,069*

No.	Built	Name	Double chimney	Withdrawn	Mileage
7023	7/49	Penrice Castle	5/58	2/65	730,636*
7024	7/49	Powis Castle	3/59	2/65	731,344*
7025	8/49	Sudeley Castle		9/64	685,916*
7026	8/49	Tenby Castle		10/64	636,668*
7027	9/49	Thornbury Castle		8/63 Preserved	728,843
7028	5/50	Cadbury Castle	10/61	12/63	624,626
7029	5/50	Clun Castle	10/59	12/65 Preserved	618,073*
7030	6/50	Cranbrook Castle	5/59	2/63	637,339
7031	6/50	Cromwell's Castle		7/63	749,715
7032	6/50	Denbigh Castle	9/60	9/64	666,374*
7033	7/50	Hartlebury Castle	7/59	2/63	605,219
7034	8/50	Ince Castle	12/59	6/65	616,584*
7035	8/50	Ogmore Castle	1/60	6/64	580,346*
7036	8/50	Taunton Castle	8/59	9/63	617,653
7037	8/50	Swindon		3/63	519,885

* Mileage to 12/63 only.
** Including mileage as 'Star' or 'The Great Bear'

BIBLIOGRAPHY

Allen, Cecil J., *Locomotives Practice & Performance*, Trains Illustrated, Ian Allan Ltd, 1960-1965
Bartlett, Steve, *Gloucester Locomotive Sheds*, Pen & Sword, 2019
Bartlett, Steve, *Hereford's 'Castle' Indian Summer*, Steam Days, March 2011
Bartlett, Steve, *Hereford's Locomotive Shed*, Pen & Sword, 2017
Bartlett, Steve, *Indian Summer of Wolverhampton (Oxley) Castles*, Steam Days, April 2015
Bartlett, Steve, *Worcester's Locomotive Shed*, Pen & Sword, 2020
Bradshaw, David, *The GWR's two million mile club*, Steam Days, April 2020
Harris, Michael, *The last days of the Castle 4-6-0s*, Steam World, February 1992
Langston, Keith, *British Steam GWR Collett Castle Class*, Pen & Sword, 2015
Leech, K. & Higson, M.F., *Pendennis Castle – 4079*, Roundhouse Books, 1965
Maidment, David, *A Privileged Journey*, Pen & Sword, 2015
Maidment, David, *An Indian Summer of Steam*, Pen & Sword, 2015
Maidment, David, *Great Western Castles, 1923-1959*, Pen & Sword, 2022
Nelson, Ronald, *Locomotive Performance, A Footplate Survey*, Ian Allan, 1979
Nock, O.S., *The GWR Stars, Castles & Kings, Vol 1*, David & Charles, 1967
Nock, O.S., *The GWR Stars, Castles & Kings, Vol 2*, David & Charles, 1970
Nock, O.S., *The Steam-hauled Bristolian*, Railway Magazine, March 1969
Penney, Derek, *GWR 4-6-0s in Colour*, Ian Allan, 1997
Reed, Brian, *Loco Profile No.3 – Great Western 4-Cylinder 4-6-0s*, Hylton Lacy Publishers Ltd, 1967
Riley R.C. & Waller, Peter, *The Power of the Castles*, Oxford Publishing Co., 2003
Rutherford, Michael, *Castles & Kings at Work*, Book Club Associates/Ian Allan Ltd, 1982
Waters, Laurence, *Castles, The Final Years 1954-1965*, Ian Allan Ltd, 2015
Waters, Laurence, *Great Western Star Class Locomotives*, Pen & Sword, 2017
Woodley, Richard, *The Day of the Holiday Express*, Ian Allan Ltd, 1996

INDEX

Bibliography, 177
Britannia Class Locomotives, 23
Castle Class Locomotives
Accidents, 110
Allocation, 23, 82-83, 85, 98, 100
Dimensions, 169
Double chimney provision, 8, 23-24
Footplate experiences, 57-59, 61, 63-74
High superheat, 8
Performance,
    Failures, 111
    Gloucester route, 38
    Highest speeds, 111
    May 9 1964, Z48 High speed finale, 89-95
    North & West route, 75-76, 85
    Oxford/Worcester route, 41-43, 60, 75, 85, 89
    South Wales route, 26-27, 33, 51, 74-75
    Various, 52-54, 110-111
    West of England route, 52
    Wolverhampton – Bristol route, 99-100
Statistics, 170-176
Tests, 8
Weight diagram, 169
Withdrawal, 82, 100, 109
Diesel Hydraulics, 9, 17, 23, 76, 85
Engineers
Churchward, G.J., 8
Collett,, C.B., 8
Ell, Sam, 8
Hawksworth, F.W., 8
King Class Locomotives, 9, 17-18
Logs

North & West route,
    4037, 57-59
Oxford/Worcester route
    5.30pm Oxford – Paddington, 40-41
    5001, 60
    7005, 47-48
Plymouth – Bristol - Paddington route
    7029, 92-93
    5054, 93-94
South Wales route
    4081, 26-27
    4099, 28-29
    5016, 28-29
    5056, 64-65
West of England route
    4079, 90-91
    7025, 91
Wolverhampton – London route
    7030, 67-73
Wolverhampton – Bristol route
    5063, 99-100
Old Oak Common Shed, 61
Photographs: Locations – Black & White
Aller Junction, 22
Barry, 10
Bristol Temple Meads, 55
Cadoxton, 11
Cardiff Canton shed, 16, 25
Cardiff Canton main line, 33, 79
Cardiff General, 13, 15, 30, 31, 34-36, 57, 80
Carmarthen shed, 88
Cholsey, 56
Craven Arms, 85
Dawlish, 14, 19

Ely Mills, 66
Evesham, 50
Exeter St David's, 21
Exminster, 20
Fenny Compton, 45
Fosse Road, 46
Gerrard's Cross, 71
Gloucester-Hereford branch, 54
Gloucester shed, 107
Gobowen, 87
Goring, 49
Hatton, 47
Hereford, 85, 87, 96-97
Hinksey, 83
Iver, 50
Laira shed, 76
Llanelli, 38, 80
Llanvihangel, 78
Miskin, 13
Nantyderry, 78
Neasden, 81
Newport, 24, 32, 37
North Pole Junction, 43
Old Oak Common, 9, 14, 73, 108
Oxford, 40, 49, 62, 81
Oxford shed, 103-104
Paddington, 35, 39, 62, 84, 89, 106
Par, 18
Pengam, 29
Pontypool Road, 55
Rattery, 19
Reading, 31, 42
Rumney River Bridge, 51
Plymouth North Road, 56, 97
Shrewsbury, 32, 59, 77
Southall, 22, 77
Southall shed, 104

Southam Road, 44
Southcote Junction, 96
Stafford Road shed, 44
Standish Junction, 101-102
Swan Village, 82
Swindon, 12, 95, 102
Swindon Works, 109
Taunton, 98
Tiverton Junction, 20
Tyseley, 108
Wantage Road, 34, 79
West Bromwich, 45
West Ealing, 39
West Wycombe, 105
Worcester shed, 85, 88, 105

Photographs: Locations – Colour
Aller Junction, 115
Ascott-under-Wychwood, 149
Berkeley Road, 156
Birmingham Snow Hill, 157
Brent, 114
Brimscombe, 148
Bristol Temple Meads, 119
Burlescombe, 116
Cardiff General, 130, 133, 140
Cardiff Canton main line, 134-135
Cardiff Canton shed, 140-142
Chipping Campden, 152
Churston, 114
Craven Arms, 146-147
Dainton, 117, 124
Dawlish, 116
Ely, 136-137
Evesham, 151
Exeter St David's, 118, 122-123
Fladbury, 148
Fratton shed, 157
Gobowen, 143
Hackney, 123
Hatton, 143-145
Hereford, 151
Honeybourne, 166
Kensal Green, 168
Laira shed, 124
Lapworth, 142
Leckwith Junction, 131, 134
Liskeard, 118

Newton Abbot, 120
North Acton, 167
Oxford, 166
Paddington, 159
Pangbourne, 129, 150, 156, 163
Plymouth Laira, 113,
Plymouth North Road, 150, 162
Reading, 125, 164-165
Shrewsbury shed, 147
Sonning, 119, 126, 155
Southcote Junction, 160
Stafford Road shed, 154
St Fagan's, 137-138
St George's, 138
Stoneycombe Quarry, 115
Swansea High St, 132, 139
Swindon Works, 153-154
Taunton, 121-122
Tilehurst, 125-127
Westbury, 161
West Ealing, 127-129, 150, 155, 158
Whiteball, 120-121
Wolverhampton Low Level, 146
Worcester shed, 153
Worcester Shrub Hill, 163
Yate, 167
Yatton, 117

**Photographs: Locomotives –
  Black & White**
Castles
100 A1, 14
111, 11
4037, 59
4074, 42, 44, 46
4077, 56
4078, 37
4079, 10, 95-96
4080, 34, 51, 78, 80
4081, 80
4082, 10
4082 (ex 7013), 44
4085, 39, 109
4087, 16, 18, 76
4089, 40
4092, 44
4096, 454097, 25
4098, 21

4099, 31
5000, 86
5005, 12
5006, 57, 109
5008, 37
5012, 79, 109
5013, 35
5014, 19
5016, 30
5021, 36
5024, 55, 109
5026, 15
5027, 108
5032, 20
5035, 109
5036, 24
5039, 29
5042, 87
5044, 36
5050, 11
5054, 88, 97-98
5055, 86
5056, 34, 56, 66, 102
5062, 32
5063, 101
5068, 46
5072, 19, 44
5076, 103
5081, 13
5085, 20, 47, 79
5090, 109
5091, 13, 77
5093, 35, 43
5095, 54, 55
5097, 14, 31
5099, 15
7001, 81
7002, 81
7005, 83
7006, 49
7007, 50, 82
7008, 45
7010, 22
7012, 77
7013 (ex 4082), 49, 62, 88
7014, 87
7016, 38

7019, 102
7020, 104
7021, 73
7022, 22
7023, 84
7024, 104
7025, 84, 97, 105
7029, 97, 105-107
7030, 71
7031, 50, 62
7032, 89
7034, 39, 107
7036, 32, 108
Other locomotives
4051, 9
4064, 12
6100, 39
6836, 19
8470, 51
70025, 79
D828, 55
D838, 86
D6327, 56
Photographs: Locomotives – Colour
Castles
4037, 116
4074, 144
4076, 127, 134
4077, 114
4079, 159-161
4080, 141-142
4081, 115
4087, 114, 117, 119, 124
4090, 132-133, 135, 143
4093, 136
4096, 122, 144
5003, 120
5011, 125
5013, 137
5015, 133
5018, 134
5021, 141
5022, 118
5032, 156
5034, 128
5037, 136
5038, 143
5039, 164
5043, 141, 146, 153
5048, 140
5050, 157
5051, 120, 135
5052, 126
5053, 117
5054, 139
5055, 121
5056, 166
5057, 157
5058, 113, 118
5059, 147
5062, 138
5063, 166
5064, 138
5065, 115
5066, 115
5071, 154
5075, 126
5078, 128
5081, 130
5089, 123
5091, 137, 140, 165
5092, 121
5093, 131
5097, 127
7000, 129, 148
7001, 146
7004, 142, 148, 154
7005, 151, 163
7007, 149-150
7008, 145, 162
7011, 156
7013 (ex 4082), 149
7015, 147
7017, 122, 155
7018, 145
7019, 167
7022, 124-125
7023, 151-153, 163
7025, 160, 165
7026, 153
7027, 152
7028, 130
7029, 158, 167-168
7031, 150
7032, 158
7034, 129, 155
7036, 119
Other locomotives
1009, 154
5563, 121
6847, 141
7205, 130
7337, 154
9752, 139
32646, 157
32650, 157
82030, 122
D800, 147
D7045, 122
Named trains, 8, 18, 26, 110
Railway Modernisation Plan, 9, 17
Rundle, Bill & Joyce, 162
Schedules
Acceleration, 8-9, 25-26, 110
Even interval, 17-18, 25
Star Class Locomotives, 9